# SpringerBriefs in Molecular Science

## Green Chemistry for Sustainability

*Series Editor*

Sanjay K. Sharma

W0080350

For further volumes:
http://www.springer.com/series/10045

Giusy Lofrano
Editor

# Emerging Compounds Removal from Wastewater

Natural and Solar Based Treatments

 Springer

Giusy Lofrano
Department of Civil Engineering
University of Salerno
Via Ponte Don Melillo 1
Fisciano 84084 Salerno
Italy

ISSN 2191-5407      e-ISSN 2191-5415
ISBN 978-94-007-3915-4      e-ISBN 978-94-007-3916-1
DOI 10.1007/978-94-007-3916-1
Springer Dordrecht Heidelberg New York London

Library of Congress Control Number: 2012933970

Printed on acid-free paper

Springer is part of Springer Science+Business Media (www.springer.com)

# Foreword

At a time when the world's population has reached seven billion people, sustainable design and environmental protection are critical to ensure that water resources will be available for future generations. It is well recognized that there is an energy/water nexus. It takes water to generate energy and energy to treat water. There is great opportunity to make wastewater treatment plants net energy users and even producers since there is 2–4 times the amount of energy embedded in wastewater than it takes to treat it. As we design wastewater treatment plants, it is important to consider the kinds of treatment that will allow us to recover energy. It is also important to recover nutrients for use as fertilizers and to reclaim water for irrigation, since there is also a water/food nexus. Technology exists which allows wastewater to be treated to a level which removes micro-contaminants such as endocrine disruptors and pharmaceuticals which not only impacts receiving waters and their uses, but also limits the ability for direct and indirect water reuse to ensure adequate supply of water. The editor, my friend and colleague Dr. Giusy Lofrano, in the framework of this book not only discusses the problems and issues associated with wastewater treatment but also offers technologically sound solutions. This book is an asset to all water professionals so they can become knowledgeable in the issues and develop sustainable design for wastewater treatment plants.

Jeanette A. Brown

# Preface

Engineering and sustainable development are intrinsically linked. Many aspects of sustainable development depend directly on appropriate and timely actions made by engineers. Green engineering focuses on how to achieve sustainability through science and technology for it is one of its fundamental principle to consider the environment when designing products and technologies.

Today, the term 'green' is used widely (and often inappropriately) in connection with many types of technologies. Generally, a technology is defined 'green' because it requires less non-renewable energy sources than others or reduces the use of hazardous chemicals. However, a truly "green" technology should consider the recycling potential, the nutrient and the energy recovery as well as ensure the preservation of ecosystems. It can be argued that green engineering is not simply good chemical engineering or industrial ecology, which alone is not enough to achieve sustainability. Indeed, even systems with efficient material and energy use can overwhelm the recovery capacity of a region or lead to other socially unacceptable outcomes.

As the quantity and quality of the resources and the resilience of the environment changes over time, the most sustainable technological solutions will change accordingly.

Green engineering was originally defined by the U.S. Congress Office of Technology Assessment as "green design involving two general goals: waste prevention and better material management". More recently, green engineering was more broadly defined by the U.S. Environmental Protection Agency (EPA) as "the design, commercialization, and use of processes and products that are feasible and economical while minimizing: generation of pollution at the source and risk to both human health and to the environment". However, sustainability is not only an issue for green engineering.

The design, the development, and the implementation of chemical products are also looking at reducing or even eliminating the use and generation of substances that may be hazardous to both human health and the environment, and therefore to green chemistry.

According to U.S. EPA, "green chemistry is required to promote innovative chemical technologies aimed at reducing or eliminating the use or generation of hazardous substances in the design, manufacture and use of chemical products." Both green engineering and green chemistry are based on twelve principles and the terms are often used interchangeably. However, although some principles may be common to both disciplines, it is clear that there are significant differences in their philosophy. Disciplines such as toxicology and thermodynamics play important roles in green chemistry despite they are not specifically included within the principles. Chapter 1 introduces green chemistry and its principles in relation to the technologies for the removal of emerging compounds from water and wastewater.

How to evaluate and to achieve sustainability in wastewater treatment plants (WWTPs)? Their crucial role in protecting human and environmental health is widely recognized. However, their impacts have simply been shifted to another part of the overall life cycle when wastewater treatment is carried out by using hazardous or non-renewable materials. Therefore, in evaluating the sustainability of WWTPs, engineers should consider the entire life cycle, including those of materials and of energy inputs. Chapter 2 reviews the removal of emerging contaminants and industrial pollutants in general from water and wastewater by using natural materials or agricultural waste as adsorbents. The problem associated with current treatment technologies lies in their lack of sustainability. If we look at centralized systems, for instance, it is clear that they are not always the best solution. The reasons are many:

- they flush contaminants out of residential areas by using large amounts of water;
- they often combine domestic wastewater with rainwater, causing the discharge of large volumes of polluted wastewater;
- they can contribute to spread a contained domestic health problem to an entire settlement or even to a region.

Furthermore, many treatment systems are functioning properly but are nevertheless unsustainable as they do not take into account the culture, the land, the climate, and the energy consumption of the country.

Chapter 3 focuses on the fate of organic chemicals in constructed wetlands and aims at improving their assessment in full-scale studies. The removal of some categories of trace contaminant of worldwide relevance, classified as Endocrine Disruptor Compounds (EDCs) and Pharmaceuticals and Personal Care Products (PPCPs), has been reviewed together with the mechanisms associated to their removal.

When a wastewater treatment technology has a high removal efficiency for contaminants, but consumes high amount of energy, this contributes to atmospheric carbon dioxide emissions. Thus, there is no net sustainability advantage in the treatment technology. In Chap. 4, the authors highlight some of the science and technology being developed to improve the solar photocatalytic decontamination of water-containing pesticides. The potential of oxidative photochemical methods using sunlight as promising alternatives to non-efficient conventional treatments is discussed in Chap. 5.

In an era when there is growing concern for the impact that our current environmental strategies has at both local and global level, it is crucial to develop more environmentally friendly wastewater treatment technologies. The hope is for these technologies to reach the environmental, economic, and societal sustainability that will contribute to reduce sanitation problems, diseased, and poverty.

Salerno, Italy, November 2011                                        Giusy Lofrano

# Acknowledgments

This book was created within the series "Springer Briefs in Green Chemistry for Sustainability" edited by Professor Sanjay Sharma. To him, my warmest thanks.

My most sincere gratitude goes to all the authors who devoted their precious time to contribute to this volume and to Sonia Ojo, Ilaria Tassistro, and their team at Springer for their valuable support that made this book possible. It has been a pleasure working with all of you.

My thanks to Jeanette Brown, who honored me by signing the Foreword, will never be enough.

I wish to express my most sincere gratitude to Süreyya Meric for all the competence showed over the years and for the scientists and friends who gave me the chance to meet.

I am beholden to Giovanni De Feo whose brilliance, wittiness and vision of the environment informed and inspired me deeply. A special thanks to Ivana Marino who supported me continuously and unconditionally and to Giovanni Pagano who helped me with his encouraging comments.

Last but not least, I am grateful to my family. Your patience and love power my life.

Salerno, Italy, October 2011                                      Giusy Lofrano

Acknowledgements

# Contents

**1 Green Chemistry for Green Treatment Technologies** .......... 1
  1.1 Introduction ....................................... 1
  1.2 Green Chemistry .................................... 3
  1.3 Green Technology ................................... 8
  1.4 Concluding Remarks ................................ 11
  References ............................................ 11

**2 Removal of Emerging Contaminants from Water
  and Wastewater by Adsorption Process** ................... 15
  2.1 Introduction ....................................... 16
  2.2 Adsorption Process ................................. 17
      2.2.1 Mechanisms and Definitions................... 17
      2.2.2 Adsorption Isotherms......................... 18
      2.2.3 Factors Affecting Adsorption ................. 19
  2.3 Removal of Emerging Compounds by Adsorption .......... 20
      2.3.1 Commercial Adsorbents........................ 21
      2.3.2 Low Cost Adsorbents ......................... 26
  2.4 Adsorption as Green Technology...................... 30
  2.5 Concluding Remarks ................................ 32
  References ............................................ 33

**3 Removal of Trace Pollutants from Wastewater
  in Constructed Wetlands** ............................... 39
  3.1 Introduction ....................................... 40
  3.2 Constructed Wetlands ............................... 40
      3.2.1 Technological Aspects........................ 40
      3.2.2 Drawbacks and Advantages .................... 42
      3.2.3 Plants Configuration ........................ 43
      3.2.4 Worldwide Diffusion......................... 44

3.3   Trace Pollutants Removal by CWs ...................... 46
      3.3.1   Endocrine Disruptors Compounds (EDCs) Removal .... 46
      3.3.2   Pharmaceuticals and Personal Care Products
              (PPCPs) Removal ........................... 48
3.4   Concluding Remarks ............................... 51
References ........................................... 52

**4   Removal of Pesticides from Water and Wastewater
     by Solar-Driven Photocatalysis** ......................... 59
4.1   Introduction ..................................... 59
4.2   Solar Photocatalysis Fundamentals. ..................... 61
4.3   Solar Photocatalysis as Green Technology ................. 62
4.4   Photocatalytic Degradation of Pesticides ................. 65
4.5   Photocatalytic Degradation of Wastewater
      Containing Pesticides............................... 70
4.6   Concluding Remarks ............................... 72
References ........................................... 74

**5   Removal of Pharmaceutics by Solar-Driven Processes** .......... 77
5.1   Introduction ..................................... 78
5.2   Solar-Based Advanced Oxidation Methods ................ 78
5.3   Treatment of Model Compounds ....................... 80
5.4   Toward Real Applications. Green Aspects of the Technology ... 84
5.5   Photolysis of Pharmaceuticals......................... 87
5.6   Concluding Remarks ............................... 88
References ........................................... 89

**6   Outlook** ........................................... 93

# Contributors

**Ana Maria Amat** Departamento de Ingeniería Textil y Papelera, Universidad Politécnica de Valencia, Campus de Alcoy, Plaza Ferrándiz y Carbonell s/n, 03801 Alcoy, Spain, e-mail: aamat@txp.upv.es

**Antonio Arques** Departamento de Ingeniería Textil y Papelera, Universidad Politécnica de Valencia, Campus de Alcoy, Plaza Ferrándiz y Carbonell s/n, 03801 Alcoy, Spain, e-mail: aarques@txp.upv.es

**Miray Bekbolet** Institute of Environmental Sciences, Bogazici University, 34342 Bebek, Istanbul, Turkey, e-mail: bekbolet@boun.edu.tr

**Vincenzo Belgiorno** Department of Civil Engineering, University of Salerno, Via Ponte don Melillo 1, 84084 Fisciano, SA, Italy, e-mail: v.belgiorno@unisa.it

**Giovanni De Feo** Department of Industrial Engineering, University of Salerno, Via Ponte don Melillo 1, 84084 Fisciano, SA, Italy, e-mail: g.defeo@unisa.it

**Mariangela Grassi** Department of Civil Engineering, University of Salerno, Via Ponte don Melillo 1, 84084 Fisciano, SA, Italy, e-mail: mgrassi@unisa.it

**Gul Kaykioglu** Faculty of Corlu Engineering, Department of Environmental Engineering, Namik Kemal University, 59860 Corlu-Tekirdag, Turkey, e-mail: gkaykioglu@nku.edu.tr

**Giusy Lofrano** Department of Civil Engineering, University of Salerno, Via Ponte don Melillo 1, 84084 Fisciano, SA, Italy, e-mail: glofrano@unisa.it

**Sixto Malato** Plataforma Solar de Almería (CIEMAT), Carretera Senés km 4, 04200 Tabernas, Almería, Spain, e-mail: Sixto.malato@psa.es

**Manuel I. Maldonado** Plataforma Solar de Almería (CIEMAT), Carretera Senés km 4, 04200 Tabernas, Almería, Spain

**Süreyya Meriç** Faculty of Corlu Engineering, Department of Environmental Engineering, Namik Kemal University, 59860 Corlu-Tekirdag, Turkey, e-mail: smeric@nku.edu.tr

**Isabel Oller** Plataforma Solar de Almería (CIEMAT), Carretera Senés km 4, 04200 Tabernas, Almería, Spain

**Günay Yıldız Töre** Faculty of Corlu Engineering, Department of Environmental Engineering, Namik Kemal University, 59860 Corlu-Tekirdag, Turkey, e-mail: gyildiztore@nku.edu.tr

**Ceyda Senem Uyguner-Demirel** Institute of Environmental Sciences, Bogazici University, 34342 Bebek, Istanbul, Turkey, e-mail: uygunerc@boun.edu.tr

**Ana Zapata** Plataforma Solar de Almería (CIEMAT), Carretera Senés km 4, 04200 Tabernas, Almería, Spain

# Abbreviations

| | |
|---|---|
| ACs | Activated carbons |
| AOPs | Advanced oxidation processes |
| AOP/BIO | Advanced Oxidation Processes/Biological Processes |
| BET | Brunauer, Emmet and Teller |
| BOD | Biochemical oxygen demand |
| BPA | Bisphenol A |
| CA | Clofibric acid |
| CC-AC | Coconut-shell based GAC |
| COD | Chemical oxygen demand |
| CWs | Constructed wetlands |
| CPCs | Compound Parabolic Concentrators |
| DBPs | Disinfection by-products |
| DOC | Dissolved organic carbon |
| DDT | Dichlorodiphenyltrichloroethane |
| DPs | Degradation products |
| EDCs | Endocrine disrupting compounds |
| EPA | Environmental Protection Agency |
| EPs | Emerging pollutants |
| FQs | Fluoroquinolones |
| FFP | Free floating plants |
| FWS | Free water surface |
| GAC | Granulated activated carbon |
| GC–MS | Gas chormatography equipped with a mass detector |
| LC–MS | Liquid chromatography equipped with a mass detector |
| LC-ESI-TOF-MS | Liquid chromatography electrospray time of flight mass spectrometry |
| HLR | Hydraulic Loading Rate |
| HPLC | High-performance liquid chromatography |
| HRT | Hydraulic Retention Time |
| HSSF-CWs | Horizontal subsurface constructed wetlands |
| ICP | Inductively coupled plasma |
| LAS | Linear alkylbenzene sulphonates |

| LECA | Light expanded clay aggregates |
| LCA | Life cycle assessment |
| MBR | Membrane biological reactor |
| NF | Nanofiltration |
| NOM | Natural organic matter |
| PAC | Powdered activated carbon |
| PAHs | Polycyclic aromatic hydrocarbons |
| PhACs | Pharmaceuticals |
| PCPs | Personal care products |
| PPCPs | Pharmaceuticals and personal care products |
| PTCs | Parabolic-trough concentrators |
| RBTSs | Reed bed treatment systems |
| RO | Reverse osmosis |
| SCI | Science Citation Index |
| SCIE | Science Citation Index Expanded |
| TOC | Total organic carbon |
| TSS | Total suspended solids |
| VSSF–CWs | Vertical subsurface constructed wetlands |
| WWTPs | Wastewater treatment plants |

# Chapter 1
# Green Chemistry for Green Treatment Technologies

Ceyda Senem Uyguner-Demirel and Miray Bekbolet

**Abstract** The implementation of clean, eco-friendly, less energy and waste producing processes and technologies is realized today with an increasing interest. In order to provide a sustainable development, environmentally friendly substances, novel technologies and new green chemistry options should be exploited. In that respect, in this chapter green chemistry and its principles are reviewed in relation to green technologies for the removal of emerging compounds from water and wastewater.

**Keywords** Oxidation processes · Green chemistry · Green technology · Treatment

## 1.1 Introduction

Population growth, global warming, resource scarcity, requirements for more energy and power bring about adverse effects of developments achieved in science and technology. The manufacture, use and disposal of chemicals obviously consume large amounts of resources, and thereby originate emissions of pollutants to all environmental compartments. With continuing innovation, each year, several hundred new chemicals are introduced while thousands of new stacks and pipes release chemical effluents into the environmental compartments i.e.; air, soil, and water.

C. S. Uyguner-Demirel (✉) · M. Bekbolet
Institute of Environmental Sciences, Bogazici University,
34342, Bebek, Istanbul, Turkey
e-mail: uygunerc@boun.edu.tr

M. Bekbolet
e-mail: bekbolet@boun.edu.tr

G. Lofrano (ed.), *Emerging Compounds Removal from Wastewater*,
SpringerBriefs in Green Chemistry for Sustainability,
DOI: 10.1007/978-94-007-3916-1_1, © Uyguner-Demirel, Bekbolet 2012

1

The global demand of chemical products will keep growing in this century, and it is expected to increase even faster than the world's population as well as gross domestic product [1]. Consequently, a shift toward a more efficient and sustainable chemistry is needed to avoid an environmental threat.

An emerging environmental philosophy first started in 1962 with the publication of Rachel Carson's "Silent Spring" which detailed the adverse effects of certain pesticides on the eggs of various birds and their spread throughout the food chain [2]. As a result of the controversy generated by this book, the use of most well-known insecticide dichlorodiphenyltrichloroethane (DDT) was banned first in the United States in 1973. Subsequently, it was also forbidden for agricultural use worldwide under the Stockholm Convention on Persistent Organic Pollutants [3], but its limited use in disease vector control continues to this day in undeveloped countries and remains controversial. In 1970, a book entitled "Since Silent Spring" [4] was published claiming that there was even more cause for alarm since "Silent Spring" was written. Another incident was the poisoning caused by consumption of contaminated fish in Minamata bay in Japan, where mercury discharged to the bay from an adjacent chemical facility was bioaccumulated and biomagnified in fish, resulting in the death of more than 100 people and the paralysis of thousands since 1956 [5].

The late sixties and early seventies were times when the environment received attention including the foundation of the Environmental Protection Agency (EPA) and the celebration of the first Earth Day on March 21, 1970. In the light of the concepts such as ecological paradigms, environmentally friendly analytical chemistry and green chemistry, advances in the improvement of new methodologies and regulatory norms that control laboratory activities can be expected in the years ahead. The literature on the subjects of clean analytical chemistry, green analytical chemistry or environmentally friendly analytical methods has grown exponentially since the 1990s due to the increasing concern of the scientific community about the environmental impact of their activity [6]. A literature survey of journals from Science Citation Index (SCI) and Science Citation Index Expanded (SCIE) database that have been published since 2000 reveals approximately 6,000 publications with keywords "green chemistry", 300 articles with keywords "green chemistry and green technology", and 22 articles when the two phrases are simultaneously searched within the titles of the publications. A number of books addressing different topics under the domain of "green chemistry" can also be attained [5, 7]. Most of them cover topics of alternative water and wastewater treatment technologies and emerging trends in environmental science [8–12]. Nameroff et al. [13] analyzed the trends and distribution of US green chemistry patents based on activity by sector and region. Moreover, only 8 articles are available in SCI-SCIE database on the topics of "green chemistry and advanced oxidation processes" all of which were published after 2006 [14, 15].

The implementation of clean, eco-friendly, less energy, and waste producing processes and technologies is realized today with an increasing interest. In order to provide a sustainable development and cope with the adverse effects of science and technology, it is a responsibility for the scientific community to work with environment friendly substances as well as novel technologies and develop new green chemistry options. In that respect, in this chapter green chemistry and its principles will be reviewed in relation to green technologies for the removal of emerging compounds from water and wastewater.

## 1.2 Green Chemistry

Green chemistry is defined as the practice of chemical science and manufacturing in a sustainable, safe, and non-polluting manner that consumes minimum amounts of materials and energy while producing little or no waste material. It can be considered as a rapidly evolving and developing subdiscipline in the field of chemistry.

Green chemistry involves a set of established principles for reducing or eliminating the use or generation of hazardous substances in the design, manufacture or application of chemical products. Moreover, it looks for alternatives on the earliest stage of materials and process design so that conventional treatment technologies can be avoided [8]. It is a highly effective approach to pollution prevention because it applies innovative scientific solutions to real-world environmental situations. Anastas and Warner [7] have given a broad definition of green chemistry based on 12 principles that relate to several steps from chemical synthesis to chemical usage describing what could be done in future for pursuing greener technologies (Table 1.1). These principles are widely accepted as a universal code of practice [16]. Considering that a green chemical should be synthesized in a safe and energy efficient manner, its toxicity should be minimal, biodegradation should be optimal and its impact to the environment should be as low as possible. In other words, it provides a road map for scientists to implement green chemistry and promotes innovation while protecting human health and the environment.

However, it is not expected that new chemical processes should always satisfy all 12 principles, but the checklist provides a rough idea of whether one process is greener than another.

The principles outlined in Table 1.1 are briefly explained and should be kept in mind in relation to topics covered throughout the manuscript. It is evident from these principles that green chemistry encompasses much more of the concepts of sustainability than simply preventing pollution.

Failure to follow the simplest rule of green chemistry (Principle 1) that is waste prevention has resulted in most of the troublesome hazardous waste sites that are causing problems throughout the world today. This was recognized in United States with the passage of the 1990 Pollution Prevention Act. This principle explicitly states that, wherever possible, wastes are not to be generated and their quantities are to be minimized. The means for accomplishing this objective can range from very simple measures, such as careful inventory control and reduction of solvent losses due to evaporation, to much more sophisticated and drastic approaches, such as complete redesign of manufacturing processes with waste

**Table 1.1** The 12 principles of green chemistry 'Reproduced from Ref. [7] with kind permission of © Oxford University Press (1998)'

Principle 1. It is better to prevent waste than to treat or clean up waste after it is formed

Principle 2. Synthetic methods should be designed to maximize the incorporation of all materials used into the final product as expressed by "atom economy"

Principle 3. Wherever practicable, synthetic methodologies should be designed to use and generate substances that possess little or no toxicity to human health and the environment

Principle 4. Chemical products should be designed to preserve efficacy of function while reducing toxicity

Principle 5. The use of auxiliary substances (e.g. solvents, separation agents) should be made unnecessary wherever possible and, innocuous when used

Principle 6. Energy requirements should be recognized for their environmental and economic impacts and should be minimized. Synthetic methods should be conducted at ambient temperature and pressure

Principle 7. A raw material of feedstock should be renewable rather than depleting wherever technically and economically practicable

Principle 8. Unnecessary derivatization (blocking group, protection/deprotection, temporary modification of physical/chemical processes) should be avoided whenever possible

Principle 9. Catalytic reagents (as selective as possible) are superior to stoichiometric reagents

Principle 10. Chemical products should be designed so that at the end of their function they do not persist in the environment and break down into innocuous degradation products

Principle 11. Analytical methodologies need to be developed to allow for real-time, in-process monitoring and control prior to the formation of hazardous substances

Principle 12. Substances and the form of a substance used in a chemical process should be chosen so as to minimize the potential for chemical accidents, including releases, explosions and fires

minimization as a top priority. One of the most effective ways to prevent generation of wastes is to ensure, as much as possible, the incorporation of all raw materials into the final product. In that respect, the concept of atom economy defined as Principle 2 is a key component of green chemistry [5]. The remaining principles are focused on issues such as toxicity, solvent and other media using consumption of energy, application of raw materials from renewable sources and degradation of chemical products to simple, nontoxic substances that are friendly for the environment.

Today, the practice of green chemistry enables designing chemicals and new approaches in a way that effectiveness is retained and even enhanced while toxicity is reduced. Chemical syntheses as well as many manufacturing operations make use of auxiliary substances that are not part of the final product. The use or generation of substances that pose hazards to humans and the environment should be minimized or totally avoided where the connection between green chemistry and environmental chemistry is especially strong. Utilization of environmentally acceptable additives for water treatment is one important area where green chemistry can potentially find applications. Alternative solvents such as supercritical fluids and ionic liquids represent another major entry in the green chemistry. The use of supercritical $CO_2$ as an environmentally friendly reaction medium for chemical synthesis, product separation and catalyst recycling has recently been revealed as an interesting clean alternative to classical organic solvents. Among various approaches, combinations of ionic liquids with supercritical fluids, particularly supercritical $CO_2$, offer a highly attractive choice [17].

On the other hand, from safety limits and regulations point of view, application of analytical techniques for the quantitative and qualitative determination of pollutants constitutes an important concern. To reduce the amount of solvent required for sample pretreatment, the application of microwave energy for sample digestion was proposed in 1975 [18]. Compared to traditional sample-preparation methods, microwave-assisted extraction saves solvent, and is rapid and efficient from the used energy point of view. Concerning the measurement step, greener analytical procedures are inherent to automated flow-based methodologies, due to their capability of reducing reagent and solvent consumption. The use of instrumental methods instead of wet chemistry, automation, and minimization is a new trend in analytical chemistry. Online classical techniques such as liquid chromatography and capillary electrophoresis have been upgraded by reducing the size of the chromatographic column, the particle size of the stationary phase or integrating the whole system on a chip. In situ monitoring allows for continuously optimizing the efficient use of reagents and permits determination of the composition of waste and effluents and their variation over time. The ability of mass spectrometry to provide fingerprints for trace level analyte is invaluable. Moreover, combinations with gas chromatography, liquid chromatography, and inductively coupled plasma (ICP) have enabled simultaneous detection and characterization of a wide range of analytes in very complex matrices. Examples of methodologies based on direct measurement are applications of spectroscopic techniques such as infrared, Raman, UV–Vis, fluorescence and nuclear magnetic resonance spectroscopy. The main advantage of these methods is to avoid sample pretreatment, thus reducing the use of solvents and reagents as well as the analysis time [6, 19]. Moreover, the recovery of reagents provides a satisfactory way to cut down the side effects of analytical methods, as it is an important step toward achieving zero emissions in research. However, there are still few of these techniques implemented in routine environmental analysis. New research in this field is now focused on improving robustness, stability, and sensitivity by using nanomaterials.

Energy consumption poses economic and environmental costs in almost all synthesis and manufacturing processes. In a broader sense, energy requirements should be minimized to avoid potential threats to the environment. One way to accomplish this goal could be the use of processes that occur at ambient conditions, rather than options requiring elevated temperature or pressure. A successful approach has been the application of biological processes for water and wastewater treatment, which, because of the conditions under which organisms grow, must occur at moderate temperatures and in the absence of toxic substances (Principle 6).

Raw materials extracted from earth are from a finite supply that cannot be replenished once they are used. So, wherever possible, renewable raw materials should be used. From the green chemistry point of view, combustion of fuels obtained from renewable feedstocks is preferred to combustion of fossil fuels from depleting finite sources. For example, many vehicles around the world are fueled with diesel oil, and the production of biodiesel oil is a promising possibility.

Hydrogen gas is used today primarily for manufacturing chemicals, but a bright future is predicted for it as a vehicle fuel in combination with fuel cells.

Synthesis of an organic compound often results in the generation of byproducts that may require disposal. Products that must be dispersed into the environment should be designed to break down rapidly into innocuous products. An example is the synthesis of biodegradable polymers [20]. Another example would be the modification of the poorly biodegradable surfactant used in household detergents. Their widespread consumption caused severe problems of foaming in wastewater treatment plants and contamination of water supplies. Hence, 15 or 20 years after their introduction, they were chemically modified to yield a product that was a biodegradable substitute.

Attaining "real-time" control of chemical processes by modern computerized systems is important for efficient and safe operation with minimum production of wastes. However, accurate knowledge on the concentrations of materials used in the system is required. Therefore, the successful practice of green chemistry requires real-time, in-process monitoring techniques (Principle 11) coupled with process control.

Accidents, such as spills, explosions, and fires, are a major hazard in the chemical industry. They are not only potentially dangerous, but also tend to spread toxic substances into the environment and increase exposure of humans and other organisms to these substances. For this reason, it is best to avoid the use or generation of substances that are likely to react violently, burn, build up excessive pressures, or otherwise cause unforeseen incidents in the manufacturing process.

The critical review of Hjeresen [10] discusses in detail green chemistry as a scientifically based set of solutions to protect water quality and prevent the growing global crisis in water resources. Warner et al. [12] address the environmental stakeholder interests in reinventing chemistry and its material inputs, products, and waste in relation to the "12 Principles of Green Chemistry". Selected examples of the implementation of green chemistry principles in everyday life in industry, laboratory, and in education were revealed in detail by Wardencki et al. [21]. Centi and Perathoner [9] introduced concepts of green chemistry with emphasis on assessment of new sustainable chemical technologies especially catalytic technologies for scaling-down chemical processes. Catalysis is a key technology to achieve the objectives of sustainable (green) chemistry, but an innovative effort is necessary in the design of new catalysts and catalytic technologies (including reactor engineering) and also in reconsidering all chemical production processes with the objective of developing small and delocalized plants for on-site production. This long-term objective requires an even more innovative effort in the direction of using catalysis in unconventional conditions. The central role of heterogeneous and homogeneous catalysis as a primary tool for achieving all of the 12 principles of green chemistry was discussed by various researchers [9, 22].

During the last two decades, with the establishment of sustainable development as a goal for society, several concepts for environmental management, that look for strategies different than just complying with environmental regulations have

**Table 1.2**  Concepts related to green chemistry [23]

| | |
|---|---|
| Green engineering | Green engineering is the design, commercialization and use of economically feasible processes and products while minimizing pollution and any threat to human and the environment |
| Cleaner production | Continuous use of an integrated and preventive environmental strategy to processes, products and services to increase the eco-efficiency and reduce risks to population and the environment |
| Eco-efficiency | It is achieved by the delivery of competitively priced goods and services that satisfy human needs and bring quality to life while reducing ecological impacts throughout life cycle in line with Earth's estimated carrying capacity |
| Industrial ecology | Examination of industry and environment where industrial system is visualized as a producer of both products and wastes and understanding the relationship between producers, consumers, other entities and the natural world |
| Ecodesign | Designing products and minimizing their direct or indirect impacts at every possible opportunity |
| Life cycle thinking | A way of addressing environmental issues and opportunities from a systematic perspective. It involves evaluating a product or service with the goal of reducing potential environmental impacts over the entire life cycle |

evolved. Aiming to achieve sustainability by introducing environmental considerations in human activities, the specific response of the chemical industry in this context has become to apply the concept of green chemistry. A definition of the most common and accepted concepts of green chemistry for environmental management as presented by Muñoz [23] are introduced in Table 1.2.

In general, these concepts aim at achieving sustainability by introducing environmental considerations in human activities. The underlying idea is the same in all of the concepts presented. Only differences on methodology, scale of application or target user can be found.

Unfortunately, the twelve principles of green chemistry do not explicitly include a number of important concepts, highly relevant to environmental impact; for example, the inherency of a product or process, the need for life cycle assessment, or the possibility of heat recovery from an exothermic reaction or heat integration. For this reason, Anastas and Zimmerman [24] subsequently proposed a set of 12 principles of green engineering (Table 1.3).

Green engineering, along with green chemistry [7], are engaged through science and technology on ensuring that quality of life, or state of economic development is increasing through benign chemicals, materials and life cycle-based design [7, 25].

The 12 Principles of green engineering (Table 1.3) provide a framework for scientists and engineers for designing effective, ecologically intelligent materials, products, and systems [26]. This approach builds on the technical excellence, scientific accuracy, and systematic thinking that have addressed in recent years the issue of science and technology for sustainability and sustainable development. Green engineering addresses the key issues at all levels of innovation.

**Table 1.3** 12 Principles of green engineering. Reprinted with permission from Anastas and Zimmerman [24]. Copyright (2003) American Chemical Society

Principle 1. Designers need to strive to ensure that all material and energy inputs and outputs are as inherently non hazardous as possible

Principle 2. It is better to prevent waste than to treat or clean up waste after it is formed

Principle 3. Separation and purification operations should be a component of the design framework

Principle 4. System components should be designed to maximize mass, energy and temporal efficiency

Principle 5. System components should be output pulled rather than input pushed through the use of energy and materials

Principle 6. Embedded entropy and complexity must be viewed as an investment when making design choices on recycle, reuse or beneficial disposition

Principle 7. Targeted durability, not immortality, should be a design goal

Principle 8. Design for unnecessary capacity or capability should be considered a design flaw which includes engineering "one size fits all" solutions

Principle 9. Multi component products should strive for material unification to promote disassembly and value retention (minimize material diversity)

Principle 10. Design of processes and systems must include integration of interconnectivity with available energy and materials flows

Principle 11. Performance metrics include designing for performance in commercial "after life"

Principle 12. Design should be based on renewable and readily available inputs throughout the life cycle

The application of the principles across disciplines has been documented in detail with case studies from a variety of sectors [24, 25].

Incorporating green engineering design principles within engineering education with specific examples for chemical engineering was revealed by Shonnard et al. [27]. The application and efficacy of green chemistry and other green design principles were documented for many case studies, including biodegradable polymers, and the production of polymers from biomaterials [28].

The need to use resources efficiently and reduce environmental impacts of industrial products and processes is becoming increasingly important in engineering design; therefore, green engineering principles are gaining prominence.

## 1.3 Green Technology

The objectives of water and wastewater treatment green technologies are: (i) to reduce and conserve the use of water and associated non-renewable energy sources; (ii) to prevent contamination and misuse of water and other natural resources; (iii) to protect biodiversity, habitats, and ecosystems, and (iv) to ensure that future generations can meet their own needs.

Taking into account current public concern on environmental matters, the consequent use of toxic reagents and solvents have increased to a point at which they became unsustainable to continue without an environmentally friendly perspective. More recently, Montreal Protocol [29] has led to the successful replacement of chlorofluorocarbons by compounds that do not affect the ozone layer appreciably. Nevertheless, many recently developed processes and products fulfill most of these principles.

The concept of embodying the chemical water treatment in green chemistry was presented by Ghernaout et al. [30] with a specific example addressing the unavoidable role of chlorination. Detailed investigation was devoted to the question of how the chemical water treatment with its chlorination in the forms of pre-disinfection, disinfection, and post-disinfection and coagulation using alum could be turned in a green water treatment.

The potential role of membranes and membrane reactors in green technologies and for water reuse were discussed in the study of Howell [31]. Examples were presented from Middle East where there is an increasing large scale use of membranes to supply potable water via reverse osmosis on brackish groundwater supplies, and for seawater desalination. It is very well established that it would be possible to reduce water use via membrane technologies such as microfiltration, ultrafiltration, nanofiltration, and reverse osmosis. With the modern developments in membrane science, fouling control is possible, membrane life times are increased and unit costs are reduced with the application of large-scale modules. Further improvement will lead to other uses of membranes that will contribute significantly to dealing with some of the major water shortage problems of the world.

Another emerging field for water and wastewater treatment is the application of advanced oxidation oxidation processes (AOPs) which are very potent in oxidation, decolorization, degradation, and mineralization of organic pollutants. Key AOPs include heterogeneous and homogeneous photocatalysis utilizing near ultraviolet or solar visible irradiation, electrolysis, ozonation, Fenton's reagent, ultrasound, and wet air oxidation, while less conventional but evolving processes include ionizing radiation, microwaves, pulsed plasma, and the ferrate reagent. The fundamental mechanisms, advantages and drawbacks, as well as the state of the art of advanced oxidation processes such as heterogeneous photocatalysis, ozonation, Fenton, and photo-Fenton processes have been documented elsewhere [32, 33]. In the study of Muñoz [23], "greenness" of these AOPs compared to other technologies for wastewater treatment were assessed using Life Cycle Assessment (LCA) as a quantitative tool. Moreover, approaches toward the development of numerous green chemical processes and wastewater treatment technologies i.e. potential applications of ozone for several types of industrial wastewaters containing recalcitrant pollutants were revealed by da Silva and Jardim [34].

Photochemistry offers numerous advantages over conventional treatment methods such as lower reaction temperatures and control of selectivity. State of the art of photochemical processes were addressed and future trends were explained in detail [35]. Implementation of UV irradiation in sample-preparation and sample-introduction systems provides remarkable improvements in analytical characteristics as well as green methods for trace-element analysis and speciation [14].

There has been an increasing interest in the publications related to photocatalytic oxidation that emphasize the process as green chemistry. Twenty five papers have been found in SCI-SCIE database most of which were published in the last 5 years. $TiO_2$ photocatalysis has shown great promise as an innovative and "green" technology due to its ability to generate electrons and holes under UV illumination, which can produce OH• radicals and initiate redox reactions to degrade trace level environmental pollutants [36, 37, 38]. The total degradation of organic pollutants such as dyes, pesticides, surfactants, phenolic compounds, aromatic, and aliphatic compounds, haloaromatics, nitrohaloaromatics, and amides that can be photodegraded using $TiO_2$ as catalyst were revealed by Blake [39]. However, the use of artificial UV light makes it difficult for this method to compete with existing ones in terms of environmental impact. In the review of Herrmann et al. [37] the photocatalytic oxidation of 4-tert-butyl-benzaldehyde was presented as an example of "green chemistry" based on the reasons that air is used and titania catalyst is a cheap, stable, and recyclable material. Moreover, the process does not require the use of solvents or heat treatment but utilizes UVA lamps the technology of which is gradually improving. It was concluded that photocatalysis addresses most of the twelve principles of "green chemistry", especially complying with the first nine principles [37, 38].

Various applications of solar photocatalysis for the decontamination of wastewater have been revealed by Robert and Malato [40]. Since then, improvement in efficiency of photocatalysis with advances in material's science, utilization of solar energy was achieved. Setting up a 'green' procedure requires that energy and reagents used at each stage should have an environmental cost with benefits. Amongst advanced oxidation processes, the applicability of photocatalysis as a technology for water and wastewater treatment is not yet successful because most data at present refer only to laboratory scale and upscaling has received only a limited attention. In the recent review of Ravelli et al. [15] environmental impact of photocatalysis was discussed. Potentially, the characteristic versatility of titania photocatalysis in the choice of conditions make the method appealing.

In the study of Mason [41] the future contribution of sonochemistry to green and sustainable science was discussed and claimed to be dependent upon the possibility of scaling up excellent laboratory results for industrial use. Some industrial scale examples from the fields of environmental protection and process technology were presented.

Electrochemistry is a rather neglected technology in the context of organic chemicals manufacturing but the green chemistry revolution opens a new door to its better exploitation. Examples of electrochemical synthesis are preparation of metal salts, in situ generation of reagents and organic electro synthesis that were described in detail in various studies [42].

As stated by Clark [43] the recipe for the twenty-first century is based on (i) designing the molecule with properties such as biodegradability and short residence time to have minimal impact on the environment, (ii) manufacturing from a renewable feedstock i.e. carbohydrate, (iii) using a long life catalyst, (iv) using no solvent or totally recyclable solvent, (v) using the smallest possible number of steps in the synthesis, (vi) manufacturing the product as required and as close as possible to where it is required.

## 1.4 Concluding Remarks

The chemical industry of $21^{st}$ century needs to fully embrace the principles of green chemistry and engineering for production of minimum waste, use of simpler and safer products, an increasing utilization of raw materials, renewable sources, and new technologies.

Green chemistry should focus in near future on development of economically feasible conversion of solar energy into chemical energy and improvement in the conversion of solar energy to electrical energy. Polluting technologies should be replaced by benign alternatives. Achieving the goals of green chemistry and green engineering involves the combined roles to be played at all aspects of society, government, and industry. Innovation and application of new cleaner technologies will probably lead to the success realization of the benefits to society and future generations.

Increasing knowledge on the production of oxidative species with higher yields, reaction pathways, reactor design, process combination, as well as applications for water reuse make AOPs a promising green treatment technology. Specifically, commercialization of photocatalysis for water and wastewater treatment could be possible on large scale in near future with the improvements in efficiency of catalyst, advances in material science and utilization of solar energy.

## References

1. OECD Environmental outlook for the chemicals industry (2001) Organization for economic co-operation and development, Paris. http://www.oecd.org/ehs
2. Carson R (1962) Silent spring. Houghton Mifflien, Boston
3. Stockholm Convention on Persistent Organic Pollutants (22 May 2001) 40 I.L.M. 532
4. Graham F Jr (1970) Since silent spring. Houghton Mifflin, Boston
5. Manahan S (2006) Green chemistry and the ten commandments of sustainability, 2nd edn. ChemChar Research, Inc Publishers, Columbia, Missouri
6. Armenta S, Garrigues S, de la Guardia M (2008) Green analytical chemistry. Trends Anal Chem 27(6):497–511
7. Anastas PC, Warner JC (1998) Green chemistry theory and practice. Oxford University Press, New York
8. Armor JN (1999) Striving for catalytically green processes in the 21st century. Appl Catal Gen 189:153–162
9. Centi G, Perathoner S (2003) Catalysis and sustainable (green) chemistry. Catal Today 77:287–297
10. Hjeresen DL (2001) Green chemistry and the global water crisis. Pure Appl Chem 73(8):1237–1241

11. Ketsetzi A, Stathoulopoulou A, Demadis KA (2008) Being "green" in chemical water treatment technologies: issues, challenges and developments. Desalination 223:487–493

12. Warner JC, Cannon AS, Dye KM (2004) Green chemistry. Environ Impact Asses Rev 24:775–799

13. Nameroff TJ, Garant RJ, Albert MB (2004) Adoption of green chemistry: an analysis based on US patents. Res Policy 33:959–974

14. Bendicho C, Pena F, Costas M, Gil S, Lavilla I (2010) Photochemistry-based sample treatments as greener approaches for trace-element analysis and speciation. Trends Anal Chem 29(7):681–191

15. Ravelli D, Fagnoni M, Dondi D, Albini A (2011) Significance of $TiO_2$ photocatalysis for green chemistry. J Adv Oxidation Technol 14(1):40–46

16. Lancaster M (2002) Principles of sustainable and green chemistry. In: Clark JH, Macquarrie D (eds) Handbook of green chemistry and technology. Blackwell Science, Oxford

17. Beckman EJ (2004) Supercritical and near-critical $CO_2$ in green chemical synthesis and processing. J Supercrit Fluids 28:121–191

18. Abu-Samra A, Morris JS, Koityohann SR (1975) Wet ashing of some biological samples in a microwave oven. Anal Chem 47:1475–1477

19. Koel M, Kaljurand M (2006) Application of the principles of green chemistry in analytical chemistry. Pure Appl Chem 78(11):1993–2002

20. Scott G (2000) "Green" polymers. Polym Degrad Stab 68(1):1–7

21. Wardencki W, Curylo J, Namieoenik J (2005) Green chemistry-current and future issues. Pol J Environ Stud 14:389–395

22. Anastas PT, Bartlett LB, Kirchhoff MM, Williamson TC (2000) The role of catalysis in the design, development, and implementation of green chemistry. Catal Today 55:11–22

23. Muñoz IO (2006) Life cycle assessment as a tool for green chemistry: application to different advanced oxidation processes for wastewater treatment, Ph.D. Thesis, Universitat Autonoma de Barcelona

24. Anastas PT, Zimmerman JB (2003) Design through the 12 principles of green engineering. Environ Sci Technol 37(5):94–101A

25. Zimmerman JB (2006) Sustainable development through the principles of green engineering, National academy of engineering: Frontiers in engineering. National Academies Press, Washington, DC

26. Mcdonough W, Braungart M, Anastas PT, Zimmerman JB (2003) Applying the principles of green engineering to cradle-to cradle design. Environ Sci Technol 1:435–441A

27. Shonnard DR, Allen DT, Nguyen M, Austin SW, Hesketh R (2003) Green engineering education through a U.S. EPA/Academia collaboration. Environ Sci Technol 37:5453–5462

28. Tabone MD, Cregg JJ, Beckman EJ, Landis AE (2010) Sustainability metrics: life cycle assessment and green design in polymers. Environ Sci Technol 44(21):8264–8269

29. Montreal Protocol on Substances that Deplete the Ozone Layer. United Nations Environment Programme (UNEP) (1987) Last amended Sept 1997. http://www.unep.ch/ozone/mont_t.htm

30. Ghernaout D, Ghernaout B, Naceur MW (2011) Embodying the chemical water treatment in the green chemistry—a review. Desalination 271:1–10

31. Howell JA (2004) Future of membranes and membrane reactors in green technologies and for water reuse. Desalination 162:1–11

32. Andreozzi R, Caprio V, Insola A, Marotta R (1999) Advanced oxidation processes (AOP) for water purification and recovery. Catal Today 53:51–59

33. Legrini O, Oliveros E, Braun AM (1993) Photochemical processes for water treatment. Chem Rev 93:671–698

34. da Silva LM, Jardim WF (2006) Trends and strategies of ozone application in environmental problems. Quim Nova 29(2):310–317

35. Dunkin IR (2002) Chapter 18. Photochemistry. In: Clark JH, Macquarrie D (eds) Handbook of green chemistry and technology. Blackwell Science Ltd, Oxford

36. Anpo M (2000) Utilization of $TiO_2$ photocatalysts in green chemistry. Pure Appl Chem 72(7):1265–1270

37. Herrmann JM, Duchamp C, Karkmaz M, Hoai BT, Lachheb H, Puzenat E, Guillard C (2007) Environmental green chemistry as defined by photocatalysis. J Hazard Mater 146:624–629
38. Herrmann JM, Lacroix M (2010) Environmental photocatalysis in action for green chemistry. Kinet Catal 51(6):793–800
39. Blake DM (1997) Bibliography of work on photocatalytic removal of hazardous compounds from water and air, NREL/TP-430-22197. National Renewable Energy Laboratory, Golden, USA
40. Robert D, Malato S (2002) Solar photocatalysis: a clean process for water detoxification. Sci Total Environ 291:85–97
41. Mason TJ (2007) Sonochemistry and the environment—providing a "green" link between chemistry, physics and engineering. Ultrason Sonochem 14:476–483
42. Clark JH, Macquarrie D (2002) Handbook of green chemistry and technology. Blackwell Science Ltd, Oxford
43. Clark JH (2002) Chapter 1 Introduction. In: Clark JH, Macquarrie D (eds) Handbook of green chemistry and technology. Blackwell Science Ltd, Oxford

# Chapter 2
# Removal of Emerging Contaminants from Water and Wastewater by Adsorption Process

Mariangela Grassi, Gul Kaykioglu, Vincenzo Belgiorno
and Giusy Lofrano

**Abstract** Emerging contaminants are chemicals recently discovered in natural streams as a result of human and industrial activities. Most of them have no regulatory standard and can potentially cause deleterious effects in aquatic life at environmentally relevant concentrations. The conventional wastewater treatment plants (WWTPs) are not always effective for the removal of these huge classes of pollutants and so further water treatments are necessary. This chapter has the aim to study the adsorption process in the removal of emerging compounds. Firstly, a brief description of adsorption mechanism is given and then the study of conventional and non-conventional adsorbents for the removal of emerging compounds is reviewed with the comparison between them.

**Keywords** Conventional adsorbents · Low-cost adsorbents · Pharmaceuticals · Personal care products · Endocrine disruptors

M. Grassi (✉) · V. Belgiorno · G. Lofrano
Department of Civil Engineering, University of Salerno,
Via Ponte don Melillo, 84084 Fisciano (SA), Italy
e-mail: mgrassi@unisa.it

V. Belgiorno
e-mail: v.belgiorno@unisa.it

G. Lofrano
e-mail: glofrano@unisa.it

G. Kaykioglu
Faculty of Corlu Engineering, Department of Environmental Engineering,
Namik Kemal University, 59860 Corlu-Tekirdag, Turkey
e-mail: gkaykioglu@nku.edu.tr

G. Lofrano (ed.), *Emerging Compounds Removal from Wastewater*,
SpringerBriefs in Green Chemistry for Sustainability,
DOI: 10.1007/978-94-007-3916-1_2, © Grassi, Kaykioglu, Belgiorno, Lofrano 2012

## 2.1 Introduction

Since the end of the last century a large amount of products, such as medicines, disinfectants, contrast media, laundry detergents, surfactants, pesticides, dyes, paints, preservatives, food additives, and personal care products, have been released by chemical and pharmaceutical industries threatening the environment and human health. Currently there is a growing awareness of the impact of these contaminants on groundwater, rivers, and lakes. Therefore the removal of emerging contaminants of concern is now as ever important in the production of safe drinking water and the environmentally responsible release of wastewater [1, 2].

Although very little investment has been made in the past on water treatment facilities, typically water supply and treatment often received more priority than wastewater collection and treatment. However, due to the trends in urban development along with rapid population increase, wastewater treatment deserves greater emphasis. Several research studies showed that, treated wastewater, if appropriately managed, is viewed as a major component of the water resources supply to meet the needs of a growing economy. The greatest challenge in implementing this strategy is the adoption of low cost wastewater treatment technologies that will maximize the efficiency of utilizing limited water resources and ensuring compliance with all health and safety standards regarding reuse of treated wastewater effluents.

Treatment options which are typically considered for the removal of emerging contaminants from drinking water as well as wastewater include adsorption, Advanced Oxidation Processes (AOPs), Nanofiltration (NF), and Reverse Osmosis (RO) membranes [3, 4]. However, the shortcomings of most of these methods are high investment and maintenance costs, secondary pollution (generation of toxic sludge, etc.) and complicated procedure involved in the treatment. On the other hand physicochemical treatments such as coagulation/flocculation processes were generally found to be unable to remove Endocrine Disrupting Compounds (EDCs) and Pharmaceuticals and Personal Care Products (PPCPs). Although AOPs can be effective for the removal of emerging compounds, these processes can lead to the formation of oxidation intermediates that are mostly unknown at this point.

Conversely adsorption processes do not add undesirable by-products and have been found to be superior to other techniques for wastewater treatment in terms of simplicity of design and operation, and insensitivity of toxic substances [5]. Among several materials used as adsorbents, Activated Carbons (ACs) have been used for the removal of different types of emerging compounds in general but their use is sometimes restricted due to high cost. Furthermore when AC has been exhausted, it can be regenerated for further use but regeneration process results in a loss of carbon and the regenerated product may have a slightly lower adsorption capacity in comparison with the virgin-activated carbon. This has resulted in attempts by various workers to prepare low cost alternative adsorbents which may replace activated carbons in pollution control through adsorption process and to overcome their economic disadvantages [6].

Recently natural materials that are available in large quantities from agricultural operations have been evaluated as low cost adsorbents and environmental friendly [7]. Moreover the utilization of these waste materials as such directly or after some minor treatment as adsorbents is becoming vital concern because they represent unused resources and cause serious disposal problems [8–11]. A growing number of studies have been carried out in recent years to evaluate the behavior of emerging adsorbents such as agricultural products and by-product for emerging contaminants removal.

On the other hand industrial wastes, such as, fly ash, blast furnace slag and sludge, black liquor lignin, red mud, and waste slurry are currently being investigated as potential adsorbents for the removal of the emerging contaminants from wastewater.

This chapter presents the state of art of wastewater treatment by adsorption focusing in special way on removal of emerging contaminants. A brief introduction of the process is first given and then the use of commercial (activated carbons, clay and minerals) and unconventional adsorbents (agricultural and industrial waste) is discussed, taking into account several criteria such as adsorption capacities ($q_e$), equilibrium time ($t_e$) and emerging contaminant removal efficiency, which make them more or less suitable to be considered green.

## 2.2 Adsorption Process

### 2.2.1 Mechanisms and Definitions

Adsorption is a mass transfer process which involves the accumulation of substances at the interface of two phases, such as, liquid–liquid, gas–liquid, gas–solid, or liquid–solid interface. The substance being adsorbed is the *adsorbate* and the adsorbing material is termed the *adsorbent*. The properties of adsorbates and adsorbents are quite specific and depend upon their constituents. The constituents of adsorbents are mainly responsible for the removal of any particular pollutants from wastewater [7].

If the interaction between the solid surface and the adsorbed molecules has a physical nature, the process is called *physisorption*. In this case, the attraction interactions are van der Waals forces and, as they are weak the process results are reversible. Furthermore, it occurs lower or close to the critical temperature of the adsorbed substance. On the other hand, if the attraction forces between adsorbed molecules and the solid surface are due to chemical bonding, the adsorption process is called *chemisorption*. Contrary to *physisorption*, *chemisorption* occurs only as a monolayer and, furthermore, substances chemisorbed on solid surface are hardly removed because of stronger forces at stake. Under favorable conditions, both processes can occur simultaneously or alternatively. Physical adsorption is accompanied by a decrease in free energy and entropy of the adsorption system and, thereby, this process is exothermic.

## 2.2.2 Adsorption Isotherms

In a solid–liquid system adsorption results in the removal of solutes from solution and their accumulation at solid surface. The solute remaining in the solution reaches a dynamic equilibrium with that adsorbed on the solid phase. The amount of adsorbate that can be taken up by an adsorbent as a function of both temperature and concentration of adsorbate, and the process, at constant temperature, can be described by an adsorption isotherm according to the general Eq. (2.1):

$$q_t = \frac{(C_0 - C_t)V}{m} \tag{2.1}$$

where $q_t$ (mg/g) is the amount of adsorbate per mass unit of adsorbent at time $t$, $C_0$ and $C_t$ (mg/L) are the initial and at time t concentration of adsorbate, respectively, $V$ is the volume of the solution (L), and $m$ is the mass of adsorbent (g).

Taking into account that adsorption process can be more complex, several adsorption isotherms were proposed. Among these the most used models to describe the process in water and wastewater applications were developed by (i) Langmuir, (ii) Brunauer, Emmet, and Teller (BET), and (iii) Freundlich.

The Langmuir adsorption model is valid for single-layer adsorption, whereas the BET model represents isotherms reflecting apparent multilayer adsorption. So, when the limit of adsorption is a monolayer, the BET isotherms reduce to the Langmuir equation. Both equations are limited by the assumption of uniform energies of adsorption on the surface.

The Langmuir isotherm is described by the Eq. (2.2):

$$\frac{q_e}{q_m} = \frac{bC_e}{1 + bC_e} \tag{2.2}$$

where $q_e$ (mg/g) is the amount of adsorbate per mass unit of adsorbent at equilibrium, $C_e$ is the liquid-phase concentration of the adsorbate at equilibrium (mg/L), $q_m$ is the maximum adsorption capacity (mg/g) and $b$ is the Langmuir constant related to the energy of adsorption (L/mg).

With the additional assumption that layers beyond the first have equal energies of adsorption, the BET equation takes the following simplified form:

$$q_e/q_m = (BC_e)/((C_S - C)[1 + (B - 1)(C_e/(C_S))]) \tag{2.3}$$

in which $C_S$ is the saturation concentration of the solute, $B$ is a constant which takes into account the energy of interaction with the surface, and all other symbols have the same significance as in Eq. (2.2).

The data related to adsorption from the liquid phase are fitted better by Freundlich isotherm equation [12]. It is a special case for heterogeneous surface energies. Freundlich isotherm is described by the Eq. (2.4):

$$q_e = K_F C_e^{1/n} \tag{2.4}$$

where $K_F$ (mg/g) (L/mg)$^{1/n}$ is the Freundlich capacity factor and $1/n$ is the Freundlich intensity parameter. The constants in the Freundlich isotherm can be determined by plotting $\log q_e$ versus $\log C_e$.

## 2.2.3 Factors Affecting Adsorption

The factors affecting the adsorption process are: (i) surface area, (ii) nature and initial concentration of adsorbate, (iii) solution pH, (iv) temperature, (v) interfering substances, and (vi) nature and dose of adsorbent.

Since adsorption is a surface phenomenon, the extent of adsorption is proportional to the specific surface area which is defined as that portion of the total surface area that is available for adsorption [13, 14]. Thus more finely divided and more porous is the solid greater is the amount of adsorption accomplished per unit weight of a solid adsorbent [15]. The major contribution to surface area is located in the pores of molecular dimensions. For example, the surface area of several activated carbon used for wastewater treatment is about 1,000 m$^2$/g, with a mean particle diameter of about 1.6 mm and density of 1.4 g/cm$^3$. Assuming spherical particles, only about 0.0003% of the total surface is the external surface of the carbon particle [16].

The physicochemical nature of the adsorbent drastically affects both rate and capacity of adsorption. The solubility of the solute greatly influences the adsorption equilibrium. In general, an inverse relationship can be expected between the extent of adsorption of a solute and its solubility in the solvent where the adsorption takes place. Molecular size is also relevant as it relates to the rate of uptake of organic solutes through the porous of the adsorbent material if the rate is controlled by intraparticle transport. In this case the reaction will generally proceed more rapidly with decrease of adsorbate molecule [15, 17–19].

The pH of the solution affects the extent of adsorption because the distribution of surface charge of the adsorbent can change (because of the composition of raw materials and the technique of activation) thus varying the extent of adsorption according to the adsorbate functional groups [15, 20–22]. For example Hamdaoui [23] showed that adsorption of methylene blue on sawdust and crushed brick increased by increasing pH (until a value of 9). For pH lower than 5 both adsorbents were positively charged: in this case, the adsorption decreased because methylene blue is a cationic dye.

Another important parameter is the temperature. Adsorption reactions are normally exothermic; thus the extent of adsorption generally increases with decreasing temperature [15, 24–26].

Finally, the adsorption can be affected by the concentration of organic and inorganic compounds. The adsorption process is strongly influenced by a mixture of many compounds which are tipically present in water and wastewater. The compounds can mutually enhance adsorption, may act relatively independently, or may interfere with one other. In most cases, as also shown hereinafter, natural organic matter (NOM) negatively affects the adsorption of emerging compounds in surface waters and wastewaters [22, 27, 28].

## 2.3 Removal of Emerging Compounds by Adsorption

Emerging contaminants are defined as compounds that are still unregulated or in process of regularization and that can be a threat to environmental ecosystems and human health [29, 30]. The words "emerging compounds" encompass a huge quantity of pollutants, including PPCPs, synthetically and naturally occurring hormones, industrial and household chemicals, nanomaterials, and some disinfection by-products (DBPs), as well as their transformation products [30]. Sources and pathways of emerging compounds into the environment depend on how (and where) they are used and how the products containing them are disposed. Figure 2.1 shows the possible contamination pathways of emerging contaminants.

The most of emerging compounds are sent to conventional Wastewater Treatment Plants (WWTPs) that allow only partial removal of micropollutants by stripping, sorption, and biological degradation.

Stripping is negligible compared with the other ones because most of emerging compounds are characterized by low volatility property. It has been demonstrated that stripping efficiency is not relevant even for musk fragrances which are slightly volatile with an Henry constant (H) value about of 0.005 [31, 32].

Sorption on primary and secondary sludge is more important than stripping process. It occurs like *absorption* on the lipid fraction of the sludge, especially on the primary sludge, and *adsorption* onto sludge through electrostatic interactions between positively charged compounds and negatively charged microorganisms surface [32]. So acid and lipophilic compounds (e.g. hormones, anti-inflammatories, fluoroquinolones) are efficiently removed in WWTPs unlike basic (clofibric acid, bezafibrate), neutral (diazepam, phenazone, and carbamazepine), and polar compounds (beta-lactam antibiotics) [33, 34].

In addition to chemical properties of specific compounds, WWTPs operating conditions are also important to study the adsorption onto sludge of emerging compounds. For example, ciprofloxacin, a polar compound, sorbed very well onto suspended solids [34], while diclofenac, which is an acid drug, is characterized from a strong variability in the removal percentage (15–80%) because of different WWTPs conditions [33, 35].

Another mechanism removal is biological degradation which is described by reaction rate constant $k_{biol}$. According to this parameter, compounds can be classified in [31, 32]:

- Highly biodegradable $k_{biol} > 10$ L/g$_{SS}$ d;
- Moderately biodegradable $0.1 < k_{biol} < 10$ L/g$_{SS}$ d;
- Hardly biodegradable $k_{biol} < 0.1$ L/g$_{SS}$ d.

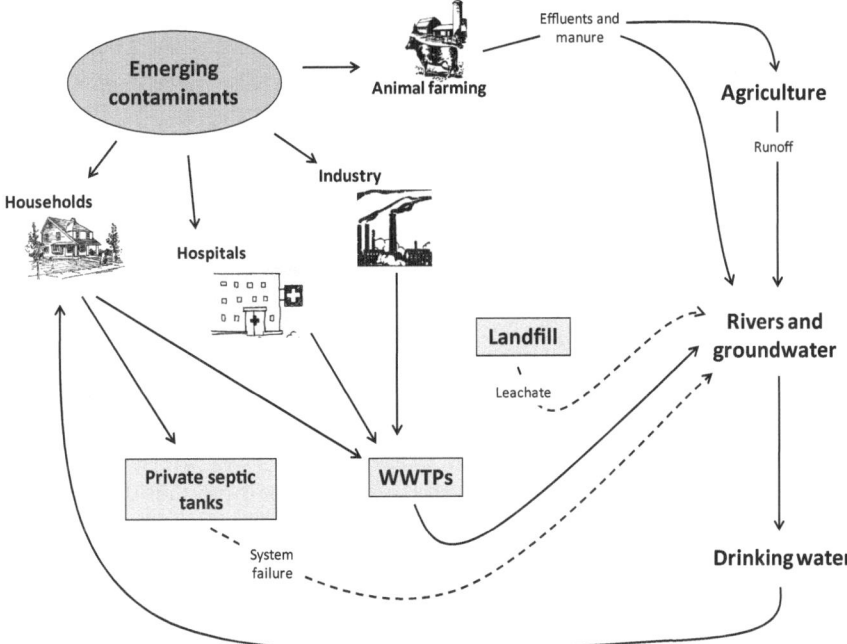

**Fig. 2.1** Potential sources and pathways of emerging compounds into the environment

Also in this case biological operating conditions are of relevant importance. Indeed, biological decomposition increases with the age of sludge [31, 34] and hydraulic retention time [36]. Some compounds are removed with low sludge age (2–5 days), other ones are hardly degradable also with sludge age greater than 20 days [34]. So, in many cases, WWTPs do not have right operating conditions to remove well-defined emerging compounds. This implies the upgrading of the plant or the use of a tertiary treatment to avoid the input of pollutants into the environment.

In the last years many studies were carried out to remove emerging pollutants by adsorption process. The most used adsorbents were commercial ones (such as natural clays, minerals, and activated carbons).

## 2.3.1 Commercial Adsorbents

### 2.3.1.1 Activated Carbon

Activated carbon prepared from different source materials (e.g. coal, coconut shells, lignite, wood, etc.) is the most popular and widely used adsorbent in wastewater treatment throughout the world. Its application in the form of carbonized wood (charcoal) has been described first in the Sanskrit medical lore and then in the Egyptian papyrus. Sanskrit writings, dating about 2,000 BC, tell how to purify impure water by boiling it in copper vessels, exposure to sunlight, and filtering through charcoal [6].

Activated carbon is produced by a process consisting of pyrolysis of raw material followed by activation with oxidizing gases. The product obtained is known as activated carbon and generally has a very porous structure with a large surface area ranging from 600 to 2,000 $m^2/g$.

Most studies concerning the removal of micropollutants in aqueous solution by adsorption are carried out by using activated carbon. However, with the aim of implementing the technology at full scale application, studies of water and waste-water are most significant. For this reason in this section only works concerning emerging contaminants found in drinking water and in wastewater will be discussed.

Redding et al. [37] evaluated the efficiency of rapid small-scale column for the treatment of a lake water spiked with 29 EDCs and PPCPs with concentration values of 100–200 ng/L. Authors studied the behavior of two kinds of carbons: a conventional activated carbon and two modified lignite carbons prepared utilizing a high-temperature steam and methane/steam. The conventional one showed a shorter bed life than modified lignite carbons. Indeed lignite variants removed EDCs/PPCPs 3–4 times longer than did commercial carbon. Furthermore the most adsorbed compounds were steroids (androstenedione, estradiol, estriol, estrone, ethynylestradiol, progesterone, and testosterone) which are characterized by quite similar molecular volume, which averaged 80 mL/mol.

The removal of 17$\beta$-estradiol from a raw drinking water was studied from Yoon et al. [22] using 5 mg/L of PAC (coal-based). The removal percentage was >90% regardless contact times and at a very low pollutant concentration (27 ng/L). This compound was also studied in the work of Yoon et al. [17]. In this study two raw drinking waters were spiked with three contaminants: 17$\beta$-estradiol, 17$\alpha$-ethynylestradiol, and bisphenol A. They were removed by adsorption on several different PAC coal-based except a wood-based one. After 1 h contact time and 45 mg/L of PAC the removal was 99% for all compounds. Increasing contact time (4 and 24 h) PAC doses were reduced (15 and 9 mg/L respectively). It is evident that contact time and adsorbent dose are important parameters in the adsorption process [28, 38]: a right combination of each allows to reach the right operating conditions in a full-scale plant.

Another important parameter is water-octanol partition coefficient (log $K_{ow}$). In particular, depending on log $K_{ow}$, hydrophobic pollutants (log $K_{ow} > 4$) have higher adsorption capacity [39–41], also if this is not always true [42, 43]. For example, Westerhoff et al. [42] evaluated the removal of 62 different EDCs/PPCPs (10–250 ng/L) in three drinking water sources. Results showed a relation between percentage EDCs/PPCPs removal and log $K_{ow}$, but not for all compounds (e.g. caffeine, pentoxyfilline). This may be related to the difficulties to accurately estimate the log $K_{ow}$ for some heterocyclic or aromatic nitrogen-containing compounds. Some results obtained by Westerhoff et al. [42] are reported in Fig. 2.2.

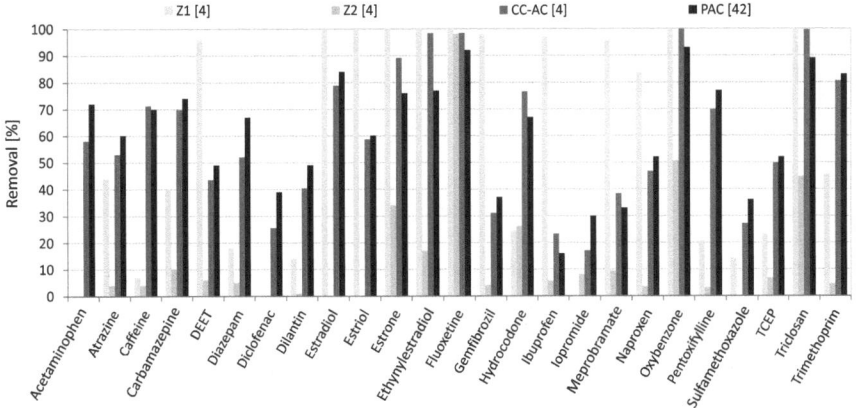

**Fig. 2.2** Removal percentages correspond to 1 mg/l dose of activated carbon CC-AC, 100 mg/l dose of Z1 and Z2 and three weeks contact time, 5 mg/l dose of PAC and 4 h contact time. Amended from Ref. [4] with kind permission of © Elsevier (2009)

As previously said, another parameter which can negatively affect adsorption process is NOM, which competes with the specific compounds for adsorption sites. It is obvious that the presence of organic matter can block pores of activated carbon and, for this reason, the removal percentage decreases if compared with results of tests carried out on model water [18, 22, 28, 44].

The problem of organic materials in water gets worse for wastewater treatment and greater carbon doses or a combination of different treatments are needed to reach a good removal percentage and to control the problem of fast deterioration of adsorbents. For instance, Hartig et al. [45] investigated the removal by PAC adsorption of two micropollutants (N–n-butylbenzenesulphonamide and sulphamethoxazole) from tertiary wastewater effluents prior to and after filtration with a tight ultrafiltration membrane. The results showed that membrane filtration prior to PAC adsorption may lead to improved elimination rates for adsorbable and low molecular weight micropollutants. Another example was reported by Baumgarten et al. [46] who examined the removal of floxacins and their precursors present in wastewater by a combination of membrane biological reactor (MBR) with PAC adsorption. PAC addition into wastewater of MBR pilot plant significantly improved removal rates (floxacins >95% and fluoroquinolonic acid as high as 77% removals at 50 mg/L initial PAC dose). Furthermore, PAC adsorption process was used to treat the permeate of MBR plant. In this case two kinds of PACs were used. The best adsorbent allowed to reach removal percentages >70% with a PAC dose of 50 mg/L. Increasing PAC dose up to 500 mg/L, a nearly complete elimination of fluoroquinolonic acids and floxacins was achieved.

The removal of micropollutants from wastewater was also carried out by the addition of commercial PAC directly to the activated sludge system with and without the adsorbent recycling to biological process [47]. Results showed that the removal efficiency increased from 30 to 50% with PAC (10 mg/L) recycling into the biological tank. Increasing PAC concentration to 15 mg/L and with PAC recycling all compounds were removed by more than 80%.

### 2.3.1.2 Clays

Natural clay minerals are well known from the earliest day of civilization. Because of their low cost, high surface area, high porosity, and abundance in most continents, clays are good candidates as adsorbents. There are many kinds of clay: smectites (montmorillonite, saponite), mica (illite), kaolinite, serpentine, pylophyllite (talc), vermiculite, sepiolite, bentonite, kaolinite, diatomite, and Fuller's earth (attapulgite and montmorillonite varieties) [6]. The adsorption capacities depend on negative charge on the surface, which gives clay the capability to adsorb positively charged species.

Putra et al. [21] investigated the removal of amoxicillin from aqueous solutions by adsorption on bentonite. A quite high value of initial amoxicillin concentration (300 mg/L) was chosen to represent pharmaceutical wastewater. Adsorption of amoxicillin was strongly affected by pH because it can alter the charge of amoxicillin molecule. In particular, $q_e$ values increased as the pH value decreased. In this study, adsorption capacity of bentonite was compared with a commercial GAC. Both adsorbents were found to be quite effective because removal percentage as high as 88% was achieved. $q_e$ value was comparable (around 20 mg/g for bentonite and 25 mg/g for commercial activated carbon), but adsorption equilibrium time for activated carbon was only 35 min compared to 8 h of bentonite. The main reason could be the different surface area of the two adsorbents: 92 m$^2$/g for bentonite and 1,093 m$^2$/g for GAC.

Bekçi et al. [24, 48] investigated montmorillonite as adsorbent in the removal of trimethoprim, one of the main antibacterial agents used in human and veterinary medicine worldwide. Results showed that the process was exothermic because of adsorption efficiency increased as temperature decreased. As a consequence of thermodynamic studies, the authors demonstrated that physisorption was the main mechanism of adsorption. Another parameter that affected adsorption of trimethoprim was pH. At low pH conditions (in an aqueous solution montmorillonite has a pH value of 3.31), trimethoprim is in the protonated form, so it was strongly adsorbed to the negatively charged surface of the montmorillonite. In the best conditions, the amount of drug adsorbed was 60 mg/g for 1 h of contact time (initial compound concentration was 290.3 mg/L).

### 2.3.1.3 Minerals

Another class of adsorbents includes natural minerals. Among these zeolite and goethite have been investigated in the adsorption of pharmaceuticals. Zeolite is typically used for the removal of dyes and heavy metals. Like clay minerals, adsorption capacity is linked to negative charge on the structure.

Ötker and Akmehmet-Balcioğlu [26] investigated the adsorption of enrofloxacin, a fluoroquinolone group antibiotic, onto natural zeolite and subsequent adsorbent regeneration by ozone treatment. The best results were achieved for lower pH values (pH investigated values were 5, 7, and 10) because of enrofloxacin is in the cationic form and so the adsorption onto negatively charged zeolite surface was better. Unlike adsorption clays, the process was endothermic, with higher enrofloxacin removal at higher investigated temperature. However, the results obtained with varying temperature (28, 37, 45, and 50 °C) showed a little change in the adsorbed amount, ranging from 16 to 18 mg/g. Adsorption equilibrium was reached at 200 min and the adsorbed amount at equilibrium was about 18 mg/g. The regeneration process by ozone oxidation (1.4 g/h) was able to decompose enrofloxacin adsorbed onto zeolite as well as to affect zeolite pore structure by decreasing pore size.

Really interesting is the study of Rossner et al. [4] concerning a lake water spiked by a mixture of 25 emerging contaminants at varying concentration (200–900 ng/L). The adsorbents used were one coconut-shell-based GAC (CC-AC), one carbonaceous resin and two high-silica zeolites, Z1 (modernite zeolite) and Z2 (Y zeolite). The order of process efficiency was activated carbon > carbonaceous resins > zeolites. Carbonaceous adsorbents were more effective for micropollutants removal probably because activated carbons exhibit a broader micropore size distribution, in which compounds of different shapes and sizes can be effectively accommodated. High-silica zeolites, on the contrary, have uniform pore sizes, which is effective for the removal of a specific compound but not for a broad mixture of contaminants.

In Fig. 2.2 results were compared with the average removal percentages obtained in four natural waters treated with 5 mg/L of powdered activated carbon (PAC) [42]. Removal values obtained with CC-AC and PAC were comparable also if the brand and the concentration of two adsorbents were different. Z1 allowed to reach high removal percentage of micropollutants but not for all compounds such as activated carbons. Z2 was the worst adsorbent and removal values were really different from Z1 (only fluoxetine, oxybenzone and triclosan were removed by Z2).

Zhang and Huang [19] investigated the removal of seven fluoroquinolones (FQs) and five structurally related model amines with Fe oxides, using two sources of goethite, with a focus on both adsorption and oxidation by Fe oxides. The authors found out that flumequine can be adsorbed more strongly to goethite than other FQs, due to effects of speciation and molecular size. Under investigated conditions (pH 5), adsorbent was positively charged, flumequine in neutral form, and the other FQs in cationic form, thus explaining the lower adsorption for the latter. Furthermore, in terms of molecular size, the other FQs being characterized

**Table 2.1** Adsorption capacity ($q_e$), initial contaminant concentration ($C_0$), and equilibrium time of some adsorbents investigated

| Adsorbent | Adsorbate | $C_0$ (mg/L) | $t_e$ (min) | $q_e$ (mg/g) | Reference |
|---|---|---|---|---|---|
| Bentonite | Amoxicillin | 300 | $\approx 500$ | 20 | [21] |
| Montmorillonite | Trimethoprim | 290.3 | 60 | 60 | [24] |
| Natural zeolite | Enrofloxacin | 200 | 200 | 18 | [26] |

by a larger molecule structure than flumequine may obstruct adsorption active sites.

In Table 2.1 are shown initial concentration of some emerging compounds and adsorption capacities (reached at a fixed equilibrium time) of some adsorbents.

## 2.3.2 Low Cost Adsorbents

Although, activated carbon is undoubtedly considered as universal adsorbent for the removal of diverse kinds of pollutants from water, its widespread use is sometimes restricted due to the high costs [8, 9, 49]. Attempts have been made to develop low-cost alternative adsorbents which may be classified in two ways (Fig. 2.3) either (i) on basis of their availability, i.e., (a) natural materials (wood, peat, coal, lignite etc.), (b) industrial/agricultural/domestic wastes or by-products (slag, sludge, fly ash, bagasse flyash, red mud etc.), and (c) synthesized products; or (ii) depending on their nature, i.e., (a) inorganic and (b) organic material [6, 8, 10, 11].

### 2.3.2.1 Agricultural Waste

The basic components of the agricultural waste materials include hemicellulose, lignin, lipids, proteins, simple sugars, water, hydrocarbons, and starch, containing a variety of functional groups [10]. In particular agricultural materials containing cellulose show a potential sorption capacity for various pollutants. If these wastes could be used as low-cost adsorbents, it will provide a two-fold advantage to environmental pollution. Firstly, the volume of waste materials could be partly reduced and secondly the low-cost adsorbent, if developed, can reduce the treatment of wastewaters at a reasonable cost [9, 50]. Agricultural waste is a rich source for activated carbon production due to its low ash content and reasonable hardness [51].

The agricultural solid wastes from cheap and readily available resources such as almond shell, hazelnut shell, poplar, walnut sawdust [52], orange peel [53, 54], sawdust [55], rice husk [56], sugarcane bagasse [57], coconut burch waste [58], and papaya seed [59] have been investigated for the removal of pollutants from aqueous solutions.

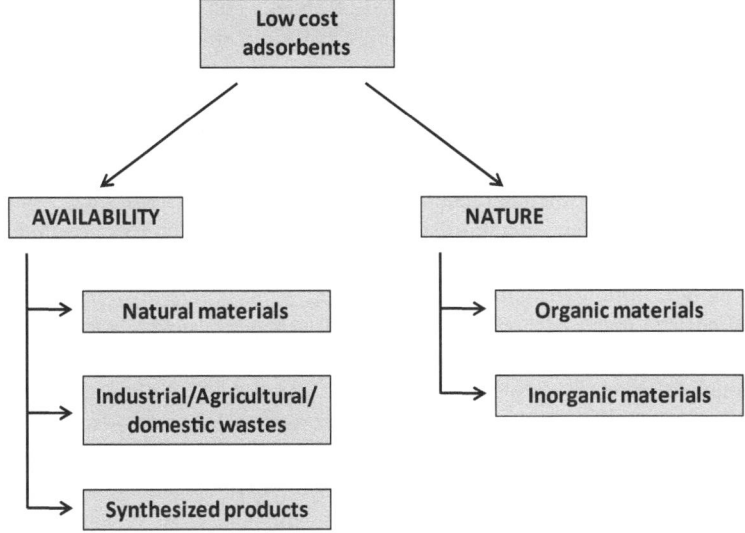

**Fig. 2.3** Possible classification of low-cost adsorbents

Sawdust [55] is one of the most appealing materials among agricultural waste materials, used for removing pollutants, such as, dyes, salts, and heavy metals from water and wastewater. The material consists of lignin, cellulose, and hemicellulose, with polyphenolic groups playing important role for binding dyes through different mechanisms. Generally the adsorption takes place by complexation, ion exchange and hydrogen bonding.

The agricultural waste materials have been used in their natural form or after some physical or chemical modification. Pretreatment methods using different kinds of modifying agents such as base solutions (sodium hydroxide, calcium hydroxide, sodium carbonate) mineral and organic acid solutions (hydrochloric acid, nitric acid, sulfuric acid, tartaric acid, citric acid), organic compounds (ethylenediamine, formaldehyde, epichlorohydrin, methanol), oxidizing agent (hydrogen peroxide), and dyes for the purpose of removing soluble organic compounds, color and metal from the aqueous solutions have been performed.

Shells of almond and hazelnut, poplar, and walnut sawdust were investigated by Aydin et al. [52] for the removal of acid green 25 and acid red 183 from aqueous solution. Equilibrium isotherms were determined and analyzed using the Freundlich equation. Capacities of adsorbent were found to be in the order: walnut > poplar >almond > hazelnut for acid green 25 and almond > walnut > poplar > hazelnut for acid red 183, respectively.

Orange peel as adsorbent has also been studied by Arami et al. [53] for the removal of direct dyes: direct red 23 and direct red 80. The authors investigated the effects of initial dye concentration (50, 75, 100, 125 mg/L), pH, mixing rate, contact time, and quantity of orange peel at 25 °C. The adsorption capacity was found to be 10.72 and 21.05 mg/g at initial pH 2 (15 min), for direct red 23 and direct red 80, respectively.

Hamdaoui [23] studied the removal of methylene blue, from aqueous solution (40 mg/L) onto cedar sawdust in order to explore their potential use as low-cost adsorbents for wastewater dye removal. Adsorption isotherms were determined at 20 °C and the experimental data obtained were modeled with the Langmuir, Freundlich, Elovich, and Temkin isotherm equations. The authors concluded that equilibrium data were well represented by a Langmuir isotherm equation with maximum adsorption capacity of 142.36 mg/g.

Rice husk as obtained from a local rice mill grounded, sieved, washed and then dried at 80 °C was used by McKay et al. [60] for removal of two basic dyes: safranine and methylene blue and adsorption capacity of 838 and 312 mg/g was found.

Batzias and Sidiras [61] studied beech saw dust as low-cost adsorbent for the removal of methylene blue and basic red 22 (1.4–14, 2.1–21 mg/L). In order to know the effect of chemical treatment and to improve its efficiency the authors also tested the potential of the adsorbent by treating it with $CaCl_2$ [61], using mild acid hydrolysis [62] and found it to increase the adsorption capacity. Further studies to evaluate the effect of pH were also carried out by Batzias et al. [63].

Shi et al. [64] improved the adsorption capacity of sunflower stalks by chemically grafting quaternary ammonium groups on them. The modified sunflower stalks exhibited increased adsorption capacity for anionic dyes, due to the existence of quaternary ammonium ions on the surface of the residues. The maximum adsorption capacities on modified sunflower stalks were found to be 191.0 and 216.0 mg/g for Congo red and direct blue, respectively, which were at least four times higher than that observed on unmodified sunflower stalks. Further, the same authors observed that adsorption rates of two direct dyestuffs were much higher on the modified residues than on unmodified ones. A comparison of various low-cost adsorbents derived from different agricultural wastes for the removal of diverse types of aquatic pollutants is summarized in Table 2.2.

### 2.3.2.2 Industrial Waste

Widespread industrial activities generate huge amount of solid waste materials as by-products. Industrial wastes such as sludge, fly ash, and red mud are classified as low-cost materials, locally available and can be used as adsorbents for removal of pollutant from aqueous solution [65].

Fly ash is a waste material originating in combustion processes. Although it may contain some hazardous substances, such as heavy metals, it has been showing good adsorption qualities for phenolic compounds [66]. The maximum phenol adsorption capacity has been found to be 27.9 mg/g for fly ash and 108.0 mg/g for granular activated carbon at initial phenol concentration of 100 mg/L.

Wang et al. [67] used fly ash as adsorbent for the removal of methylene blue from aqueous solution reporting an adsorption capacity of 4.47 mg/g. The effect

Table 2.2 Adsorption capacities ($q_e$), initial contaminant concentration ($C_0$), and equilibrium time of different agricultural wastes for removal of pollutants from aqueous solutions

| Adsorbent | Adsorbate | $C_0$ (mg/L) | $t_e$ (min) | $q_e$ (mg/g) | Reference |
|---|---|---|---|---|---|
| Rice husk (water washed) | Cd(II) | 50 | 600 | 8.58 | [68] |
| Rice husk (sodium hydroxide) | | 50 | 240 | 20.24 | |
| Rice husk (sodium bicarbonate) | | 50 | 60 | 16.18 | |
| Rice husk (epichlorohydrin) | | 50 | 120 | 11.12 | |
| Rice husk ash | Methylene blue | | | 690 | [69] |
| Jack fruit peel | Methylene blue | 50 | 180 | 285.713 | [70] |
| Papaya seed | Methylene blue | 50 | 180 | 555.557 | [71] |
| Bael fruit shell (Ortho-phosphoric acid) | Cr (VI) | 75 | 240 | 17.27 | [72] |
| Tea waste | Cu | 100 | 90 | 48 | [73] |
| | Pb | | | 65 | |
| Hazelnut shell | Ni (II) | | | 10.11 | [74] |
| Sugarcane bagasse | PAHs | 15 | | 0.345 | [75] |
| Green coconut shells | | 15 | | 0.553 | |
| Orange peel | Direct red 23 | 50 | 15 | 10.72 | [53] |
| | Direct red 80 | 50 | 15 | 21.05 | |
| Rice husk | Safranine | | | 838 | [60] |
| | Methylene blue | 312 | 360 | 312 | |
| Cedar sawdust | Methylene blue | 40 | | 142.36 | [23] |
| Beech sawdust | Methylene blue | 14 | | 9.78 | [61] |
| | Basic red 22 | 21 | | 20.2 | |
| Beech sawdust ($H_2SO_4$) | Methylene blue | 14 | | 30.5 | [62] |
| | Basic red 22 | 21 | | 24.10 | |

of physical (heat) and chemical treatment was also studied on as-received fly ash. The heat treatment was reported to have adverse effect on the adsorption capacity of fly ash but acid treatment (by nitric acid) resulted in an increase of adsorption capacity of fly ash (7.99 mg/g).

Bhatnagar and Jain [9] investigated steel and fertilizer industries wastes, as an adsorbent for the adsorption of cationic dyes. It was found that the adsorbents prepared from blast furnace sludge, dust and slag have poor porosity and low surface area, resulting in very low efficiency for adsorption of dyes.

Smith et al. [76] reported that chemical activation using alkali metal hydroxide reagents, especially KOH, was found to be the most effective technique for producing high BET surface area sludge-based adsorbents (in excess of 1,800 $m^2$/g).

Red mud is a waste material formed during the production of alumina [77]. Red mud has been explored as an alternate adsorbent for arsenic. An alkaline aqueous medium (pH 9.5) favored As(III) removal, whereas the acidic pH range (1.1–3.2) was effective for As(V) removal [78, 79]. A comparison of various low-cost adsorbents derived from different industrial wastes for the removal of diverse types of aquatic pollutants is summarized in Table 2.3.

## 2.4 Adsorption as Green Technology

The literature studies showed above highlighted that adsorption process can be considered an efficient treatment for the removal of emerging compounds from water. It allows to reach good removal percentage and, furthermore, being a physical process, does not imply by-products formation, which could be more toxic than parent compounds. It is obvious that adsorption process is encompassed in an integrated treatment system which involves many factors, such as available space for the construction of treatment facilities, waste disposal constraints, desired finished water quality, and capital and operating costs. All these factors imply the achievement of the optimal operating conditions for low-cost high efficiencies [10, 80].

The most used and studied adsorbents are certainly activated carbons both for synthetic and real water (surface water and wastewater). In spite of large use of them, the overall idea is to reduce the use of activated carbon because of high costs. Therefore, scientific world is looking for low-cost adsorbents for water pollution. In addition to cost problem, another important factor pushing toward low-cost adsorbents is the use of agricultural and industrial waste products in order to extend the life of waste materials without introducing into the environment new materials as adsorbents and to reduce costs for waste disposal therefore contributing to environmental protection. Anyway a suitable non-conventional low-cost adsorbent should:

(1) be efficient to remove many and different contaminants,
(2) have high adsorption capacity and rate of adsorption, and
(3) have high selectivity for different concentrations.

**Table 2.3** Adsorption capacities ($q_e$), initial contaminant concentration ($C_0$), and equilibrium time of different industrial wastes for removal of pollutants from aqueous solutions

| Adsorbent | Adsorbate | $C_0$ (mg/L) | $t_e$ (min) | $q_e$ (mg/g) | Reference |
|---|---|---|---|---|---|
| Carbon slurry of fertilizer industry | Ethyl orange | $5.10^{-4}$M | 45 | 198 | [50] |
| | Metanil yellow | $5.10^{-4}$M | 45 | 211 | |
| | Acid blue 113 | $5.10^{-4}$M | 45 | 219 | |
| Exhausted olive cake ash | Ni (II) | 100 | 120 | 8.38 | [81] |
| | Cd (II) | 100 | 120 | 7.32 | |
| Clarified sludge | Cr(VI) | 50 | 120 | 26.31 | [82] |
| Activated alumina | | 50 | 180 | 25.57 | |
| Fly ash | | 50 | 180 | 23.86 | |
| Raw fly ash | Methylene blue | | | 4.47 | [67] |
| Fly ash (acid treatment) | | | | 7.99 | |
| Fly ash | As (V) | 50 | 480 | 19.46 | [83] |
| Fertilizer industrial waste (carbon slurry waste) | 4-bromophenol | $4.10^{-4}$M | 480 | 40.7 | [84] |
| | 2-bromophenol | $4.10^{-4}$M | 480 | 170.4 | |
| | 2.4-dibromophenol | $4.10^{-4}$M | 480 | 190.2 | |
| Modified basic oxygen furnace slag | Reactive blue 19 | 100 | 180 | 76 | [85] |
| | Reactive black 5 | 100 | 180 | 60 | |
| | Reactive red 120 | 100 | 180 | 55 | |

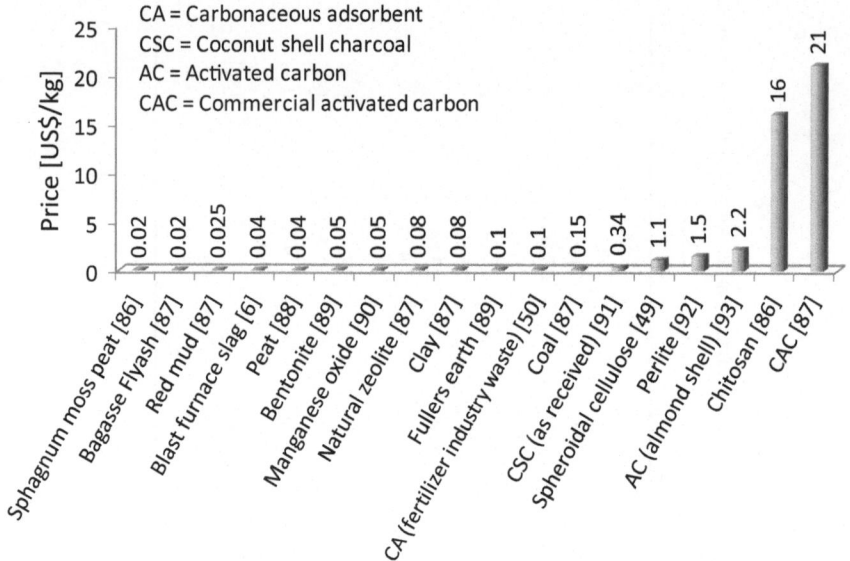

**Fig. 2.4** Cost of several adsorbents. The image contains the references to the respective absorbents in square brackets [6, 49, 50, 86–93]

It is very difficult to understand which adsorbent is better because they have different properties (porosity, surface area, and physical strength) as well as different adsorption capacities related to experimental conditions [94].

Adsorbent cost is an important parameter to compare different materials. In Fig. 2.4 costs of several low-cost and commercial adsorbents are shown.

They should be considered indicative because of adsorbent costs depend on many factors such as its availability, its source (natural, industrial/agricultural/domestic wastes or by-products or synthesized products), treatment conditions, and recycle and lifetime issues. Furthermore, the cost also depends on when adsorbents are produced in (or for) developed, developing, or underdeveloped countries [95]. Finally, a right cost evaluation is related to the application scale and, although many studies about non-conventional low-cost adsorbents are available in the literature, they are limited to laboratory scale. Thus, cost estimation is not strictly right and pilot-plant studies should also be conducted utilizing low-cost adsorbents to check their feasibility on commercial scale.

## 2.5 Concluding Remarks

The economical and easily available adsorbent would certainly make an adsorption-based process a viable alternative for the treatment of wastewater containing pollutants. Selection of an appropriate adsorbent is one of the key issues to achieve the maximum removal of type of pollutant depending upon the adsorbent and adsorbate characteristics. The effectiveness of the treatment depends not only on the properties of the adsorbent and adsorbate, but also on various environmental conditions and variables used for the adsorption process, e.g. pH, ionic strength, temperature, existence of competing organic or inorganic compounds in solution, initial adsorbate and adsorbent concentration, contact time and speed of rotation, particle size of adsorbent, etc.

# References

1. Kümmerer K (2009) The presence of pharmaceuticals in the environment due to human use – present knowledge and future challenges. J Environ Manage 90:2354–2366
2. Zuccato E, Castiglioni S, Fanelli R, Bagnati R (2007) Inquinamento da farmaci: le evidenze (parte I). Ricerca&Pratica 23:67–73
3. Bolong N, Ismail AF, Salim MR, Matsuura T (2009) A review of the effects of emerging contaminants in wastewater and options for their removal. Desalination 239:229–246
4. Rossner A, Snyder SA, Knappe DRU (2009) Removal of emerging contaminants of concern by alternative adsorbents. Water Res 43:3787–3796
5. Tong DS, Zhou CH, Lu Y, Yu H, Zhang GF, Yu WH (2010) Adsorption of acid red G dye on octadecyl trimethylammonium montmorillonite. Appl Clay Sci 50:427–431
6. Gupta VK, Carrott PJM, Ribeiro Carrott MML, Suhas TL (2009) Low-cost adsorbents: growing approach to wastewater treatment—a review. Crit Rev Env Sci Technol 39:783–842
7. Khattri SD, Singh MK (2009) Removal of malachite green from dye wastewater using neem sawdust by adsorption. J Hazard Mater 167:1089–1094
8. Ahmaruzzaman MD (2008) Adsorption of phenolic compounds on low-cost adsorbents: A review. Adv Colloid Interf Sci 143:48–67
9. Bhatnagar A, Jain AK (2005) A comparative adsorption study with different industrial wastes as adsorbents for the removal of cationic dyes from water. J Colloid Interf Sci 281:49–55
10. Bhatnagar A, Sillanpää M (2010) Utilization of agro-industrial and municipal waste materials as potential adsorbents for water treatment—a review. Chem Eng J 157:277–296
11. Wan Ngah WS, Hanafiah MAKM (2008) Removal of heavy metal ions from wastewater by chemically modified plant wastes as adsorbents: a review. Bioresour Technol 99:3935–3948
12. Cooney DO (1999) Adsorption design for wastewater treatment. CRC Press LLC, Boca Raton
13. El-Sheikh AH, Newman AP, Al-Daffaee H, Phull S, Cresswell N, York S (2004) Deposition of anatase on the surface of activated carbon. Sur Coat Technol 187:284–292
14. Naeem A, Westerhoff P, Mustafa S (2007) Vanadium removal by metal (hydr)oxide adsorbents. Water Res 41:1596–1602
15. Weber WJJr (1972) Physicochemical processes for water quality control. Wiley, New York
16. Culp RL, Wesner GM, Culp GL (1978) Handbook of advanced wastewater treatment. Van Nostrand Reinholds Company, New York
17. Yoon Y, Westerhoff P, Snyder SA, Esparza M (2003) HPLC-fluorescence detection and adsorption of bisphenol A, 17β-estradiol, and 17α-ethynyl estradiol on powdered activated carbon. Water Res 37:3530–3537
18. Yu Z, Peldszus S, Huck PM (2009) Adsorption of selected pharmaceuticals and an endocrine disrupting compound by granular activated carbon. 2. Model prediction. Environ Sci Technol 43:1474–1479
19. Zhang H, Huang CH (2007) Adsorption and oxidation of fluoroquinolone antibacterial agents and structurally related amines with goethite. Chemosphere 66:1502–1512
20. Gao J, Pedersen JA (2005) Adsorption of Sulfonamide Antimicrobial Agents to Clay Minerals. Environ Sci Technol 39:9509–9516

21. Putra EK, Pranowo R, Sunarso J, Indraswati N, Ismadji S (2009) Performance of activated carbon and bentonite for adsorption of amoxicillin from wastewater: Mechanisms, isotherms and kinetics. Water Res 43:2419–2430
22. Yoon Y, Westerhoff P, Snyder SA (2005) Adsorption of $^3$H-labeled 17-$\beta$estradiol on powdered activated carbon. Water Air Soil Pollut 166:343–351
23. Hamdaoui O (2006) Batch study of liquid-phase adsorption of methylene blue using cedar sawdust and crushed brick. J Hazard Mater B135:264–273
24. Bekçi Z, Seki Y, Yurdakoç MK (2006) Equilibrium studies for trimethoprim adsorption on montmorillonite KSF. J Hazard Mater B133:233–242
25. Önal Y, Akmil-Başar C, Sarici-Özdemir Ç (2007) Elucidation of the naproxen sodium adsorption onto activated carbon prepared from waste apricot: Kinetic, equilibrium and thermodynamic characterization. J Hazard Mater 148:727–734
26. Ötker HM, Akmehmet-Balcioğlu I (2005) Adsorption and degradation of enrofloxacin, a veterinaryantibiotic on natural zeolite. J Hazard Mater 122:251–258
27. Saravia F, Frimmel FH (2008) Role of NOM in the performance of adsorption-membrane hybrid systems applied for the removal of pharmaceuticals. Desalination 224:168–171
28. Snyder SA, Adham S, Redding AM, Cannon FS, DeCarolis J, Oppenheimer J, Wert EC, Yoon Y (2007) Role of membranes and activated carbon in the removal of endocrine disruptors and pharmaceuticals. Desalination 202:156–181
29. Esplugas S, Bila DM, Krause LGT, Dezotti M (2007) Ozonation and advanced oxidation technologies to remove endocrine disrupting chemicals (EDCs) and pharmaceuticals and personal care products (PPCPs) in water effluents. J Hazard Mater 149:631–642
30. La Farré M, Pérez S, Kantiani L, Barcelo D (2008) Fate and toxicity of emerging pollutants, their metabolites and transformation products in the aquatic environment. Trends Anal Chem 27(11):991–1007
31. Joss A, Zabczynski S, Gobel A, Hoffman B, Lffler D, McArdell CS, Ternes TA, Thomsen A, Siegrist H (2006) Biological degradation of pharmaceuticals in municipal wastewater treatment: proposing a classification scheme. Water Res 40:1686–1696
32. Suarez S, Lema JM, Omil F (2010) Removal of pharmaceutical and personal care products (PPCPs) under nitrifying and denitrifying conditions. Water Res 44:3214–3224
33. Beausse J (2004) Selected drugs in solid matrices: a review of environmental determination, occurrence and properties of principal substances. Trends Anal Chem 23:753–761
34. Ternes TA, Joss A, Siegrist H (2004) The complexity of these hazards should not be underestimated. Environ Sci Technol 38:392–399A
35. Paxeus N (2004) Removal of selected non-steroidal anti-inflammatory drugs (NSAIDs), gemfibrozil, carbamazepine, $\beta$-blockers, trimethoprim and triclosan in conventional wastewater treatment plants in five EU countries and their discharge to the aquatic environment. Water Sci Technol 50:253–260
36. Auriol M, Filali-Menassi Y, Tyagi RD, Adams CD, Surampalli RY (2006) Endocrine disrupting compounds removal from wastewater, a new challenge. Process Biochem 41:525–539
37. Redding AM, Cannon FS, Snyder SA, Vanderford BJ (2009) A QSAR-like analysis of the adsorption of endocrine disrupting compounds, pharmacuticals, and personal care products on modified activated carbons. Water Res 43:3849–3861
38. Tanghe T, Verstraete W (2001) Adsorption of nonylphenol onto granular activated carbon. Water Air Soil Pollut 131:61–72
39. Bertanza G, Pedrazzani R, Zambarda V (2009) I microinquinanti organici nelle acque di scarico urbane: presenza e rimozione. Ingegneria Ambientale 48
40. Choi KJ, Kim SG, Kim CW, Kim SH (2005) Effects of activated carbon types and service life on removal of endocrine disrupting chemicals: amitrol, nonylphenol, and bisphenol-A. Chemosphere 58:1535–1545
41. Stackelberg PE, Gibs J, Furlong ET, Meyer MT, Zaugg SD, Lippincott RL (2007) Efficiency of conventional drinking-water-treatment processes in removal of pharmaceuticals and other organic compounds. Sci Total Environ 377:255–272

42. Westerhoff P, Yoon Y, Snyder S, Wert E (2005) Fate of endocrine-disruptor, pharmaceutical, and personal care product chemicals during simulated drinking water treatment processes. Environ Sci Technol 39:6649–6663

43. Yu Z, Peldszus S, Huck PM (2008) Adsorption characteristics of selected pharmaceuticals and endocrine disrupting compound—naproxen, carbamazepine and nonylphenol—on activated carbon. Water Res 42:2873–2882

44. Yu Z, Peldszus S, Huck PM (2009) Adsorption of selected pharmaceuticals and an endocrine disrupting compound by granular activated carbon. 1. adsorption capacity and kinetics. Environ Sci Technol 43:1467–1473

45. Hartig C, Ernst M, Jekel M (2001) Membrane filtration of two sulphonamides in tertiary effluents and subsequent adsorption on activated carbon. Water Res 35(16):3998–4003

46. Baumgarten S, Schröder HFr, Charwath C, Lange M, Beier S, Pinnekamp J (2007) Evaluation of advanced treatment technologies for the elimination of pharmaceutical compounds. Water Sci Technol 56(5):1–8

47. Zwickenpflug B, Boehler M, Dorusch F, Hollender J, Fink G, Ternes T, Siegrist H (2010) International symposium "20 years of research in the field of endocrine disruptors & pharmaceutical compounds", Berlin 10 Feb 2010

48. Bekçi Z, Seki Y, Yurdakoç MK (2007) A study of equilibrium and FTIR, SEM/EDS analysis of trimethoprim adsorption onto K10. J Mol Struct 827:67–74

49. Babel S, Kurniawan TA (2003) Low-cost adsorbents for heavy metals uptake from contaminated water: a review. J Hazard Mater 97:219–243

50. Jain AK, Gupta VK, Bhatnagar A, Suhas TL (2003) Utilization of industrial waste products as adsorbents for the removal of dyes, J Hazard Mater 101:31–42

51. Ahmedna M, Marshall WE, Rao RM (2000) Production of granular activated carbons from selected agricultural by-products and evaluation of their physical, chemical and adsorption properties. Bioresour Technol 71:113–123

52. Aydin AH, Bulut Y, Yavuz O (2004) Acid dyes removal using low cost adsorbents. Int J Environ Pollut 21:97–104

53. Arami M, Limaee NY, Mahmoodi NM, Tabrizi NS (2005) Removal of dyes from colored textile wastewater by orange peel adsorbent: equilibrium and kinetic studies. J Colloid Interf Sci 288:371–376

54. Namasivayam C, Muniasamy N, Gayatri K, Rani M, Ranganathan K (1996) Removal of dyes from aqueous solutions by cellulosic waste orange peel. Bioresour Technol 57:37–43

55. Shukla A, Zhang Y-H, Dubey P, Margrave JL, Shukla SS (2002) The role of sawdust in the removal of unwanted materials from water. J Hazard Mater 95:137–152

56. Vadivelan V, Kumar KV (2005) Equilibrium, kinetics, mechanism, and process design for the sorption of methylene blue onto rice husk. J Colloid Interf Sci 286:90–100

57. Ibrahim SC, Hanafiah MAKM, Yahya MZA (2006) Removal of cadmium from aqueous solution by adsorption on sugarcane bagasse. Am-Euras. J Agric Environ Sci 1:179–184

58. Hameed BH, Mahmoud DK, Ahmad AL (2008) Equilibrium modeling and kinetic studies on the adsorption of basic dye by a low-cost adsorbent: Coconut (Cocos nucifera) bunch waste. J Hazard Mater 158:65–72

59. Hameed BH (2009) Removal of cationic dye from aqueous solution using jackfruit peel as non-conventional low-cost adsorbent. J Hazard Mater 162:344–350

60. McKay G, Porter JF, Prasad GR (1999) The removal of dye colours from aqueous solutions by adsorption on low-cost materials. Water Air Soil Pollut 114:423–438

61. Batzias FA, Sidiras DK (2004) Dye adsorption by calcium chloride treated beech sawdust in batch and fixed-bed systems. J Hazard Mater 114:167–174

62. Batzias FA, Sidiras DK (2007) Dye adsorption by prehydrolysed beech sawdust in batch and fixed-bed systems. Bioresour Technol 98:1208–1217

63. Batzias FA, Sidiras DK (2007) Simulation of dye adsorption by beech sawdust as affected by pH. J Hazard Mater 141:668–679

64. Shi WX, Xu XJ, Sun G (1999) Chemically modified sunflower stalks as adsorbents for color removal from textile wastewater. J Appl Polym Sci 71:1841–1850

65. Gulnaz O, Kaya A, Matyar F, Arikan B (2004) Sorption of basic dyes from aqueous solution by activated sludge. J Hazard Mater 108:183–188
66. Aksu Z, Yener J (1999) The usage of dried activated sludge and fly ash wastes in phenol biosorption/adsorption: comparison with granular activated carbon. J Environ Sci Health Part A 34:1777–1796
67. Wang S, Boyjoo Y, Choueib AA (2005) Comparative study of dye removal using fly ash treated by different methods. Chemosphere 60:1401–1407
68. Kumar U, Bandyopadhyay M (2006) Sorption of cadmium from aqueous solution using pretreated rice husk. Bioresour Technol 97:104–109
69. Chandrasekhar S, Pramada PN (2006) Rice husk ash as an adsorbent for methylene blue-effect of ashing temperature. Adsorption 12:27–43
70. Hameed BH (2009) Removal of cationic dye from aqueous solution using jackfruit peel as non-conventional low-cost adsorbent. J Hazard Mater 162:344–350
71. Hameed BH (2009) Spent tea leaves: a new non-conventional and low-cost adsorbent for removal of basic dye from aqueous solutions. J Hazard Mater 161:753–759
72. Anandkumar J, Mandal B (2009) Removal of Cr(VI) from aqueous solution using Bael fruit (Aegle marmelos correa) shell as an adsorbent. J Hazard Mater 168:633–640
73. Amarasinghe BMWPK, Williams RA (2007) Tea waste as a low cost adsorbent for the removal of Cu and Pb from wastewater. Chem Eng J 132:299–309
74. Demirbas O, Alkan M, Dogan M (2002) The removal of Victoria blue from aqueous solution by adsorption onto low-cost material. Adsorption 8:341–349
75. Crisafully R, Milhome MAL, Cavalcante RM, Silveira ER, De Keukeleire D, Nascimento RF (2008) Removal of some polycyclic aromatic hydrocarbons from petrochemical wastewater using low-cost adsorbents of natural origin. BioresourTechnol 99:4515–4519
76. Smith KM, Fowler GD, Pullket S, Graham NJD (2009) Sewage sludge-based adsorbents: a review of their production, properties and use in water treatment applications. Water Res 43:2569–2594
77. Mohan D, Pittman CU Jr (2007) Review Arsenic removal from water/wastewater using adsorbents—a critical review. J Hazard Mater 142:1–53
78. Altundogan HS, Altundogan S, Tumen F, Bildik M (2000) Arsenic removal from aqueous solutions by adsorption on red mud. Waste Manag 20(8):761–767
79. Altundogan HS, Altundogan S, Tumen F, Bildik M (2002) Arsenic adsorption from aqueous solutions by activated red mud. Waste Manag 22:357–363
80. Oller I, Malato S, Sanchez-Pérez JA (2011) Combination of advanced oxidation processes and biological treatments for wastewater decontamination—a review. Sci Total Environ 409:4141–4166
81. Elouear Z, Bouzid J, Boujelben N, Feki M, Montiel A (2008) The use of exhausted olive cake ash (EOCA) as a low cost adsorbent for the removal of toxic metal ions from aqueous solutions. Fuel 87:2582–2589
82. Bhattacharya AK, Naiya TK, Mandal SN, Das SK (2008) Adsorption, kinetics and equilibrium studies on removal of Cr(VI) from aqueous solutions using different low-cost adsorbents. Chem Eng J 137:529–541
83. Li Y, Zhang F-S, Xiu F-R (2009) Arsenic(V) removal from aqueous system using adsorbent developed from a high iron-containing fly ash. Sci Total Environ 407:5780–5786
84. Bhatnagar A (2007) Removal of bromophenols from water using industrial wastes as low cost adsorbents. J Hazard Mater 139:93–102
85. Xue Y, Hou H, Zhu S (2009) Adsorption removal of reactive dyes from aqueous solution by modified basic oxygen furnace slag: isotherm and kinetic study. Chem Eng J 147:272–279
86. Sharma DC, Forster CF (1993) Removal of hexavalent chromium using sphagnum moss peat. Water Res 27:1201–1208
87. Lin SH, Juang RS (2009) Adsorption of phenol and its derivatives from water using synthetic resins and low-cost natural adsorbents: a review. J Environ Manage 90:1336–1349
88. USGS(a) (2005) minerals yearbook: Peat. http://minerals.usgs.gov/minerals/pubs/commodity/peat/peat myb05.pdf. Accessed 30 June 2011

89. USGS(b) (2007). 2005 minerals handbook: Clay. http://minerals.usgs.gov/minerals/pubs/commodity/clays/claysmyb05.pdf. Accessed 30 June 2011

90. Chakravarty S, Dureja V, Bhattacharyya G, Maity S, Bhattacharjee S (2002) Removal of arsenic from groundwater using low cost ferruginous manganese ore. Water Res 36:625–632

91. Babel S, Kurniawan TA (2004) Cr(VI) removal from synthetic wastewater using coconut shell charcoal and commercial activated carbon modified with oxidizing agents and/or chitosan. Chemosphere 54:951–967

92. Mathialagan T, Viraraghavan T (2002) Adsorption of cadmium from aqueous solutions by perlite. J Hazard Mater 94:291–303

93. Toles CA, Marshall WE, Wartelle LH, McAloon A (2000) Steam- or carbon dioxide-activated carbons from almond shells: physical, chemical and adsorptive properties and estimated cost of production. Bioresour Technol 75:197–203

94. Crini G (2006) Non-conventional low-cost adsorbents for dye removal: A review. Bioresour Technol 97:1061–1085

95. Gupta VK, Suhas TL (2009) Application of low-cost adsorbents for dye removal—A review. J Environ Manag 90:2313–2342

# Chapter 3
# Removal of Trace Pollutants from Wastewater in Constructed Wetlands

**Günay Yıldız Töre, Süreyya Meriç, Giusy Lofrano and Giovanni De Feo**

**Abstract** The first experiments using constructed wetland for wastewater treatment were carried out in Germany in the early 1950s. Since then, their potential for removal conventional contaminants from wastewater is well established, making of them a technology suitable to fulfill important remediation strategies. Furthermore recent studies assessed the ability of CWs to remove trace pollutants. This chapter focuses on the fate of trace pollutants in constructed wetlands and aims at improving their assessment in full-scale studies. The removal of some categories of trace contaminant of worldwide relevance, classified as endocrine disruptors compound (EDCs) as well as pharmaceuticals and personal care products (PPCPs), has been reviewed together with mechanisms associated to their removal.

**Keywords** Constructed wetlands · EDCs · Pharmaceuticals · Personal care products · Trace pollutants

G. Y. Töre · S. Meriç
Çorlu Engineering Faculty, Environmental Engineering Department,
Namık Kemal University, Çorlu, Tekirdağ, Turkey
e-mail: gyildiztore@nku.edu.tr

S. Meriç
e-mail: smeric@nku.edu.tr

G. Lofrano (✉)
Department of Civil Engineering, University of Salerno,
via Ponte don Melillo 1, 84084, Fisciano (SA), Italy
e-mail: glofrano@unisa.it

G. De Feo
Department of Industrial Engineering, University of Salerno,
via Ponte don Melillo 1, 84084, Fisciano (SA), Italy
e-mail: g.defeo@unisa.it

G. Lofrano (ed.), *Emerging Compounds Removal from Wastewater*,
SpringerBriefs in Green Chemistry for Sustainability,
DOI: 10.1007/978-94-007-3916-1_3, © Töre, Meriç, Lofrano, De Feo 2012

## 3.1 Introduction

During the last decade, the occurrence of organic micropollutants into the environment gained growing interest since a generalized concern arises about the possible undesirable effects of many of these contaminants on human health [1]. These trace pollutants, usually known as emerging pollutants, mainly consist of compounds of anthropogenic origin such as pharmaceutical (PhACs) and personal care products (PCPs), pesticides, surfactants and plasticizers that are continuously discharged into the environment as a result of consumer activities, waste disposal, accidental releases and purposeful introduction [2–4]. Most of these organic pollutants are only partially eliminated in conventional wastewater treatment plants (WWTPs). Therefore one of the main sources of these pollutants into the environment is the discharge of effluents from WWTPs, where they have been detected in concentrations ranging from ng/1 to low µg/1 [5, 6]. Despite their low concentrations, their ecotoxicological effects are unpredictable because of the large number of compounds possibly present and their design as biologically active molecules [7].

In order to decrease the load of organic pollutant discharge into the environment, a number of technologies have been attempted, as shown in the other chapters of this book, however the potential use of constructed wetlands has been only partially explored.

Up till now, due to the high surface/equivalent-inhabitant ratio required to achieve wastewater quality parameters, CWs are only feasible in small communities or as tertiary treatments dealing with a small, diverted fraction of conventional WWTPs effluents [5]. However, both alternatives are attractive because CWs show a high capacity to remove organic micropollutants, particularly pharmaceutical and personal care products (PPCPs), consuming few energy and with relatively low maintenance costs, representing a green technology indeed.

The chapter introduces the technological aspects of constructed wetlands, discussing drawbacks and advantages. Successively it summarizes the results of literature studies on removal of trace pollutants from wastewater in constructed wetlands, together with recent progresses made toward understanding the mechanism attributed to organic chemicals removal.

## 3.2 Constructed Wetlands

### 3.2.1 Technological Aspects

Constructed wetlands (CWs) are engineered systems that have been designed and constructed to reproduce the processes occurring in natural wetland within a more controlled environment. Wastewater treatment is achieved through an integrated combination of biological, physical and chemical interactions among plants, substrate and soil [8, 9].

The soil is the main supporting material for plant growth and microbial films. Moreover, the soil matrix has a decisive influence on the hydraulic processes. Both chemical soil composition and physical parameters such as grain-size distributions, interstitial pore spaces, effective grain sizes, degrees of irregularity and the coefficient of permeability are all important factors influencing the biotreatment system [9].

The main mechanisms of contaminants removal from wastewater in constructed wetlands are microbial processes such as nitrification and denitrification for nitrogen as well as physicochemical processes such as the fixation of phosphate by iron and aluminum in the soil filter. Moreover, plant nutrient uptake and tissue storage of nutrients as well as heavy metals play an important role in the treatment processes plants. The role of macrophytes in constructed treatment wetlands is acknowledged by several authors [8–10]. Fifteen years ago Brix [8] firstly suggested that as wetlands plants are very productive, considerable amounts of nutrients can be bound in the biomass. Since that time, metabolic transformations of different organic chemicals have been shown to occur in a variety of plants [11, 12], including typical constructed wetland plants like the common reed (*P. australis*), the broad-leaved cattail (*Typha latifolia*) and some popular species (*Populus p.*) [13, 14]. The extent to which plants can degrade organic chemicals mainly depends on the specific compound of interest [9].

According to the life form of the dominating macrophyte, CWs may be classified into systems with free-floating, floating leaved, rooted emergent and submerged macrophytes [15].

Further division could be made according to the wetland hydrology (free water surface and subsurface systems) and subsurface flow CWs could be classified according to the flow direction [16] as vertical or horizontal (Fig. 3.1). In surface-flow wetlands (FWS) the wastewater flows through a shallow basin planted with emergent and submerged macrophytes. This kind of system is mainly exploited for tertiary treatment or polishing stage and also in several cases of diffuse pollution. In subsurface flow or "Reed-bed" treatment systems (RBTS), the wetland is filled with gravel or sand or similar substrates, and the plants, most commonly Reeds (*P. australis* or *communis*), grow rooted in the filling medium. The direction of the water flow provides the names of the two most diffused designs for RBTSs, the horizontal flow (HSSF-CWs) and vertical flow (VSSF-CWs) systems.

The first experiments aimed at the possibility of wastewater treatment by wetland plants were undertaken by Käthe Seidel in Germany in the early 1950s at the Max Planck Institute in Plön [17]. Seidel then carried out numerous experiments aimed at the use of wetland plants for treatment of various types of wastewater, including phenol wastewaters [18], dairy wastewaters [19] or livestock wastewater [20].

Most of her experiments were carried out in constructed wetlands with either horizontal subsurface (HSSF-CWs) or vertical subsurface (VSSF-CWs) flow, but the first fully constructed wetland was built with free water surface (FWS) in the Netherlands in 1967 [21]. However, FWS CWs did not spread substantially in Europe where subsurface flow constructed wetlands prevailed in the 1980s and 1990s [16].

**Fig. 3.1** Constructed
wetlands for wastewater
treatment. **a** CW with free-
floating plants (FFP). **b** CW
with free water surface and
emergent macrophytes
(FWS). **c** CW with horizontal
subsurface flow (HSSF, HF).
**d** CW with vertical
subsurface flow (VSSF, VF)

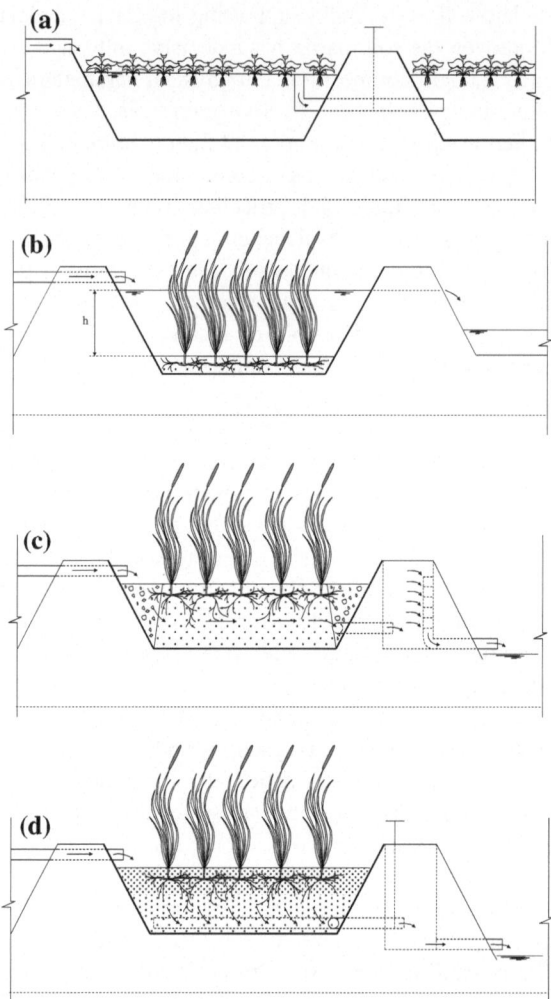

### 3.2.2 Drawbacks and Advantages

After five decades of research and implementation, CWs have been recognized
as a reliable wastewater treatment technology and, at present, they represent a
suitable solution for treatment of many types of wastewater. Most former
concerns regarding their safe and reliable application have been refuted
[16, 22, 23].

Compared with other treatment technologies, CWs present several advantages
such as low required energy input, low operational cost, and simple operation
and maintenance (O/M), which make them particularly suitable for wastewater

**Table 3.1** Comparison of removal efficiency of some parameters for common biological treatment systems

| Treatment systems | Removal efficiency (%) | | | |
|---|---|---|---|---|
| | BOD | N | P | Coliforms |
| SP | 75–90[a] | 30–50[a] | 20–60[a] | 60–90[a] |
| LFRT | 85–90[a] | 30–40[a] | 30–45[a] | 60–90[a] |
| AT | 60–80[a] | 10–25[a] | 10–20[a] | 60–90[a] |
| AS | 85–95[a] | 30–40[a] | 30–35[a] | 60–90[a] |
| CWs | 60–85[b] | 79–94[e*] | 28–41[c] | >95[d] |

*SP* Stabilization ponds, *LRTF* Low rate trickling filters, *AT* Anaerobic treatment, *AS* Activated sludge, *CW* constructed wetlands-emergent macrophytes

[a] [28]
[b] [32]
[c] [33]
[d] [30] ($10^2 - 10^3$ CFU/100 ml final level)
[e] [6][*] ($N-NH_4$)

treatment in urban and rural areas [24–29]. The low cost, easy construction (no advanced technology needed) and operation (no chemical use, no qualified personnel requirement) as well as high performance for removal of conventional and toxic pollutants and pathogens [30, 31] make constructed wetlands preferable to other treatment options as shown in Table 3.1

On the other hand several reports on CWs performance in different locations have noted several disadvantages [34–41]. First, studies have shown that the worst problem for CW is the progressive clogging that occurs near the inlet, resulting from solids entrapment and sedimentation, biofilm growth, plant decay products, granular medium properties and chemical precipitation [34, 38, 39, 42]. One significant factor influencing the clogging process is total suspended solid (TSS) load in wastewater [36, 41, 43]. Second, temperature influences the performance of constructed wetlands, especially in winter. Phosphate removal efficiency decreases rapidly with decreasing temperature. Third, CWs also have relatively low or unstable performance in the start-up period due to immature rhizosphere environments. Finally, CWs need large areas of land and a longer retention time to achieve acceptable effluent water quality. Thus, pre-treatment technologies for constructed wetlands are needed to polish and lower the pollution load [35, 44, 45].

## 3.2.3 Plants Configuration

Wastewater treatment technologies, such as septic tanks (or Imhoff tanks) for small-scale installations and primary decanters for larger installations have been used in CWs pre-treatment. When septic tanks are properly operated, TSS removal efficiency is only 50–70%, and phosphate and organic matter removal is also very limited [46–48]. The worst problem in using primary decanters is the large amount of primary sludge produced [49].

Compared with the above conventional technologies, chemical coagulation pre-treatment for CWs has some advantages [50–52]. First, through the coagulation flocculation processes, the added chemicals agglomerate the suspended solid particles and increase their sedimentation rate [53, 54].

This treatment allows higher removal efficiency of TSS, and lower chemical oxygen demand (COD), phosphorus and turbidity [35, 50, 51, 53, 54]. For example, studies have shown that the coagulation flocculation process allows the removal of 80–90% of TSS and 40–70% of BOD [49, 55]. This demonstrates potential to reduce the contaminant load to CWs. Second, chemical coagulation as a CWs pre-treatment has been shown to require only half the sedimentation pond volume of other conventional methods [56].

Constructed wetlands could be combined in order to achieve a higher treatment effect by using advantages of individual systems. Most hybrid constructed wet-lands combine VSSF and HSSF stages but, in general, all types of constructed wetlands could be combined [44, 45].

The VSSF-HSSF system was originally designed by Seidel as early as in the late 1950s and the early 1960s [57] but the use of hybrid systems was then very limited. In the 1980 s VSSF-HSSF hybrid constructed wetlands were built in France [58] and United Kingdom. At present, hybrid constructed wetlands are in operation in many countries around the world and they are used especially when removal of ammonia-N and total-N is required [16].

Besides sewage, hybrid constructed wetlands have been used to treat a variety of other wastewaters, for example, landfill leachate [59, 60], compost leaching [61], slaughterhouse [62], shrimp and fish aquaculture [63, 64] or winery [65].

### 3.2.4 Worldwide Diffusion

The widespread diffusion of constructed wetlands in projects of sustainable wastewater management for wastewater reclamation and reuse relies on their compliance with water quality guidelines to minimize human and ecosystem health risks [30, 31].

India is one of the first developing countries where a full-scale constructed wetland was installed. The plant was found to be efficient in removal of BOD and N with low costs and low energy requirements [33]. Constructed wetland have been commonly used even in countries with high population densities, such as Denmark or The Netherlands [10, 66].

In North America, FWS CWs started with the ecological engineering of natural wetlands for wastewater treatment at the end of the 1960s and beginning of the

**Table 3.2** Examples of the use of HSSF-CWs for various types of industrial wastewater ([96]—reprinted from Ref. [48] with kind permission of the American Chemical Society 2002)

| Type of wastewater | Location | References |
|---|---|---|
| Petrochemical | USA, China | [97, 98] |
| Chemical industry | United Kingdom | [99] |
| Paper and pulp wastewaters | USA | [100] |
| Abattoir (Slaugheterhouse) | Mexico, Ecuador | [101, 102] |
| Textile industry | Australia | [103] |
| Tannery industry | Portugal | [104] |
| Food industry | Slovenia, Italy | [105, 106] |
| Distillery and winery | India, Italy | [65, 107] |
| Pig farm | Australia, Lithuania | [108, 109] |
| Fish farm | Canada, Germany | [110, 111] |
| Dairy | US, Germany, Uruguay | [112–114] |

All references cited are reported in Vymazal [96]

1970s [67–69]. This treatment technology was adopted in North America not only for municipal wastewaters but all kinds of wastewaters [22]. Subsurface flow technology spread more slowly in North America but, at present, thousands of CWs of this type are in operation [15].

In the Mediterranean basin, very successful experiences with CWs have been reported for France [70–73], Spain [74], Portugal [75], Morocco [76], Italy [65, 77–80], Egypt [25, 81], Israel [82], Slovenia [83, 84], Croatia [85], Greece [86], Turkey [87].

The CWs have been used for different kind of wastewaters such as municipal and domestic wastewater gray water, rain water, landfill as well as industrial and agricultural wastewater, urban and highway run-off [8, 31, 88].

When the efficiency of constructed wetlands to transform and recycle nutrients from wastewater is well known [8, 89–91]. It is not necessary to achieve high removal of nutrients for the purpose of effluent reuse for irrigation such as before to discharge in water bodies [92].

Regarding pathogen removal, the constructed wetlands are able to furnish an effluent corresponding to B category which could be reused for irrigation of crops, fodder and trees [93–95]. The high removal efficiency of pathogens which allows achieving fecal coliform concentrations between 100 CFU/100 ml and 1000 CFU/100 ml [30, 31] is compatible with a wide range of reuse applications including irrigation.

Although it is currently being researched, there is still no consensus regarding the threshold for PPCPs in reclaimed water reuse.

Tables 3.2 and 3.3 reports examples of the use of subsurface horizontal flow systems (HSSF-CWs) and FWS for various type of industrial wastewater.

**Table 3.3** Examples of the use of FWS CWs for various types of industrial wastewater ([96]—Reprinted from Ref. [48] with kind permission of the American Chemical Society 2002)

| Type of wastewater | Location | References |
|---|---|---|
| Animal wastes | USA | [115–117] |
| Dairy pasture runoff | New Zealand | [118] |
| Acid coal mine drainage | USA, Spain | [32, 119] |
| Metal ores mine drainage | Germany, Ireland, Canada | [120–122] |
| Refinery process waters | USA, Hungary | [123, 124] |
| Paper and pulp wastewaters | USA | [125] |
| Sugar factory | Kenya | [126] |
| Olive mill | Greece | [127] |
| Metallurgic industry | Argentina | [128, 129] |

All references cited are reported in Vymazal [96]

## 3.3 Trace Pollutants Removal by CWs

### 3.3.1 Endocrine Disruptors Compounds (EDCs) Removal

According to COM (1999) 706 "Community Strategy for Endocrine Disrupters", a potential endocrine disrupter is an exogenous substance or mixture that possesses properties that might be expected to lead to endocrine disruption in an intact organism, or its progeny or sub population. Their class includes phthalates, alchiphenol ethoxylates, phenols and phenols compounds, natural and synthetically produced hormones, estrogens, polychlorobiphenils (PCB), pesticides, polycyclic aromatic hydrocarbons (PAHs). Nowadays there are still no reuse standards for municipal sewage containing hazardous substances such as xenobiotic or endocrine disrupters (EDCs) in many countries.

The behavior of some of them has been studied in constructed wetlands pointing out that reed beds are effective in the removal of phthalates, alkylphenol ethoxylates, estrogens, PAHs and several types of pesticides [1–3, 130, 131]. The removal of linear alkylbenzene sulphonates (LAS) was also investigated [96, 130, 132, 133]. LAS are the most widely used synthetic anionic surfactants. Due to their frequent use in laundry and cleaning products, LAS are a common constituent of municipal and industrial wastewaters. Table 3.4 reports range of removal efficiency of some EDCs in constructed wetlands.

According to Table 3.4, a relatively large variation in the wastewater concentration of PAHs can be observed among the studies of Masi et al. [131] and Fontoulakis et al. [130]. However it is almost common for PAHs because urban wastewater receives deposits of PAHs from different sources such as car traffic, industries, waste incinerators and domestic heating via both atmospheric transport and local activity [130]. It is worth to notice that independently from the initial concentration, the removal of PAHs was similarly high. According to these studies the main removal mechanisms for PAHs and LAS seem to be adsorption in solid media and secondly biodegradation. Whereas the removal of PAHs increased with

**Table 3.4** Range of removal efficiency of some EDCs in CWs

| Parameter | Values | Unit | Removal (%) |
|---|---|---|---|
| PAHs | 15–180 | ng/l | 60–70[a] |
|  | 786 ± 514 | ng/l | 68–79[b] |
| Steroid estrogens | 164–259 | ng/l | 100[a] |
| Estrone | 0.39–10.49 | ng/l | 67.8 ± 28[c] |
| 17 beta-estradiol | 1.35–9.05 | ng/l | 84 ± 15.4[c] |
| 17 alpha-ethynylestradiol | 0.59–6.56 | ng/l | 75.3 ± 17.6[c] |
| Diethylphthalate | 151–3,788 | ng/l | 80–100[a] |
| Di-n-butylphatalate | 43–6,134 | ng/l | 100[a] |
| Bisphenol A | 0.05–0.3 | μg/l | 80–100[d] |
| Linear alkylbenzene sulphonates (LAS) | 1.2–17.2 | mg/l | 30–55[b] |

[a] [131]
[b] [130]
[c] [3]
[d] [1]

mass loading rate in subsurface constructed wetlands and decreased in FWS, LAS were poorly removed in FWS even at low or high mass loading rates [130]. Studying the removal of bisphenol A (BPA), a phenolic estrogenic compound in a pilot scale horizontal subsurface flow constructed wetlands, Avila et al. [1] suggested the biodegradation and association to the particulate matter as the most likely processes involved in the elimination of BPA. The removal of estrone, 17 beta-estradiol and 17 alpha-ethynylestradiol in three CWs with different filter layer depth (7.5, 30 and 60 cm) was investigated by Song et al. [3]. Together with the result that the performance of wetlands when operating in unsaturated condition was superior to that when operating in water-saturated condition, the authors observed that maintaining sufficient aerobic circumstance in constructed wetlands was important for estrogens removal. The highest efficiency of estrogen removal was achieved in extremely shallow wetland might be due partly to the highest root density, besides the superior condition for penetration of oxygen. The adsorbed estrogens in sand accounted for less than 12% of the removed estrogens irrespective of the depth, indicating biotic processes play a major role in the estrogens removal.

Constructed wetlands have also been evaluated to treat wastewater contaminated with pesticides.

Agudelo et al. [2] reported the results obtained from a 6 months run study with simultaneous removal of chlorpyrifos (used in agriculture to prevent and control pests, cattle parasites and as a pesticide for a wide variety of crops) and dissolved organic matter in water using four horizontal subsurface flow constructed wetlands (HSSF-CW) at a pilot scale, that were planted with *P. australis*, at 20 ± 2° C water temperature. The removals were assumed possibly due to mineralization processes, biological decomposition and sorption in plants.

Table 3.5 reports removal efficiency in constructed wetlands of some common used pesticides.

**Table 3.5** Range of removal efficiency of some pesticides in CWs (SF surface CW–SSF subsurface CW)

| Pesticides | Concentration (µg/l) | Removal (%) | |
|---|---|---|---|
| | | SF | SSF |
| Chlorpyrifos | 268, 49[a] | | >96[a] |
| Mecoprop | 7.80 ± 3.24[b] | 79–91[b] | 22[c] |
| MCPA | 2.01 ± 1.50[b] | 79–93[b] | |
| Tertbutulazine | 2.30 ± 1.82[b] | 80[b] | |

[a] [2]
[b] [6]
[c] [134]

## 3.3.2 Pharmaceuticals and Personal Care Products (PPCPs) Removal

PPCPs constitute a large and diverse group of organic compounds used throughout the world. These substances are excreted by humans in varying degrees of degradation and discharged directly into the sewerage systems and wastewater treatment plants. The world wide occurrence of PPCPs such as hormones, nonsteroidal antiinflammatory drugs, fragrance, lipid regulators, beta blockers and psychiatric drugs has been recently reported by Fatta-Kassinos et al. [4]. There has been an intensive effort to study the factors such as plant type, design criteria, hydraulic regimes, carrier materials affecting the removal mechanisms of specific PPCPs in constructed wetlands in different design and application. Table 3.6 reports range of removal efficiency in constructed wetlands of some PPCS.

Conkle et al. [136] showed that the percent reductions of pharmaceuticals (PhACs) observed in the Mandeville lagoon constructed wetlands system were greater than reduction rates reported for conventional WWTPs; perhaps due to the longer treatment time (>>30 days). Most target PhACs were not completely removed before discharge into Lake Pontchartrain, although their collective annual loading was reduced to less than 1 kg and down to ppb with significant potential for dilution in the large lake.

The sorption capacity of light expanded clay aggregates (LECA) was evaluated to remove mixtures of ibuprofen, carbamazepine, considered one of the most recalcitrant pharmaceuticals and clofibric acid, a metabolite of blood lipid regulator drugs, in view of using these materials in CWs [137]. High removal efficiencies were attained for carbamazepine and ibuprofen while a less satisfactory performance was observed for clofibric acid. In a mixture of the three compounds in water a slight decrease in the sorbed amounts was observed in comparison with solutions of the single compounds, indicating some competitive sorption. When two other clay materials—sepiolite and vermiculite—were tested for the removal of the more recalcitrant clofibric acid, vermiculite exhibited higher removal efficiency than LECA. A study by Dordio et al. [138] was also conducted to assess *Typha* spp.'s ability to withstand and remove, from water clofibric acid (CA). At a

**Table 3.6** Range of removal efficiency of some PPCPs in CWs (SF surface CW—SSF subsurface CW)

| PPCPs | Use | Concentration ($\mu$g/l) | Removal (%) | |
|---|---|---|---|---|
| | | | SF | SSF |
| Ibuprofen | Analgesic | $0.04 \pm 0.03^{a,b,c}$ | – | 71[a] |
| | | | $96 \pm 2^b$ | – |
| | | | 95–96[c] | – |
| | | 1.5–56.5[d] | – | 50–100[d] |
| | | 4.3–7.3[e] | – | 71–79.7[e] |
| Naproxen | Analgesic | $0.34 \pm 0.06^{a,b,c}$ | – | 85[a] |
| | | | 52–92[b] | – |
| | | | $72 \pm 28^c$ | – |
| | | 0.3–2.2[d] | – | 80–100[d] |
| | | 2.6–4.3[e] | – | 82.8–91.3[e] |
| Diclofenac | Analgesic | $1.25 \pm 0.11^{a,b,c}$ | – | n.r[a] |
| | | | 73–96[b] | – |
| | | | $85 \pm 16^c$ | – |
| | | 0.003–0.3[d] | – | 80–100[d] |
| | | 11.3–12.8[e] | – | 47–55[e] |
| Ketoprofen | Analgesic | $2.10–0.70^{a,b,c}$ | – | n.r[a] |
| | | | 97–99[b] | – |
| | | | $98 \pm 1^c$ | – |
| Clofibric acid | Antilipidic | $0.07 \pm 0.01^{a,b,c}$ | – | n.r[a] |
| | | | 32–36[b] | – |
| | | | $34 \pm 3^c$ | – |
| Carbamazepine | Antiepileptic | $0.37 \pm 0.08^{a,b,c}$ | – | 16[a] |
| | | | 30–37[b] | – |
| | | | $39 \pm 12^c$ | – |
| | | 17–17.9[e] | – | 26.7–28.4[e] |
| Tonalide | Fragrance | $0.86 \pm 0.10^{a,b,c}$ | – | 88[a] |
| | | | 88–90[b] | – |
| | | | $89 \pm 1^c$ | – |
| | | 0.04–0.1[d] | – | 80–100[d] |
| Galaxolide | Fragrance | $2.86 \pm 0.10^{a,b,c}$ | – | 86 |
| | | – | 88–90[b] | – |
| | | – | $87 \pm 2^c$ | – |

[a] [134]
[b] [6]
[c] [31]
[d] [1]
[e] [135]

concentration of 20 $\mu$g/L, *Typha* had removed >50% of CA within the first 48 h, reaching a maximum of 80% by the end of the assay.

A continuous injection experiment was implemented in a pilot-scale horizontal subsurface flow constructed wetland system to evaluate the behavior of four pharmaceuticals and personal care products (i.e. ibuprofen, naproxen, diclofenac and tonalide) and a phenolic estrogenic compound [1]. Naproxen and diclofenac were efficiently removed (93%) after preanaerobic degradation. Tonalide was readily removed in the small wetlands where the removal of total suspended solids was 93%. Given its high hydrophobicity, sorption onto the particulate matter stands for the major removal mechanism.

Horizontal subsurface flow constructed wetlands were evaluated to remove carbamazepine, ibuprofen and clofibric acid by Matamoros et al. [5]. The removal efficiencies were found to be independent of the organic matter type (i.e. dissolved or particulate). Carbamazepine was removed inefficiently (5%) by bed sorption, whereas ibuprofen was removed by degradation (51%). In addition, the behavior of the two main ibuprofen biodegradation intermediates (carboxy and hydroxy derivatives) supported that the main ibuprofen elimination pathway occurs in aerobic conditions. A further study carried out by the same authors [6], analysing both the dissolved and particulate phases from the influent and effluent of a subsurface CWs, confirmed that whereas a major proportion of emerging pollutants occurred in the dissolved phase, galaxolide and tonalide were strongly associated with the particulate phase due to their high hydrophobicity (log Kow ¼ 5.7–5.9).

The ability of tropical horizontal subsurface constructed wetlands (HSSF-CWs) planted with *Typha angustifolia* to remove four widely used pharmaceutical compounds (carbamazepine, declofenac, ibuprofen and naproxen) at the relatively short hydraulic residence time of 2–4 days was documented by Zhang et al. [135] too. For both ibuprofen and naproxen, pharmaceutical compounds with low Dow values, the planted beds showed significant ($p < 0.05$) enhancement of removal efficiencies (80 and 91%, respectively, at the 4 day HRT), compared to unplanted beds (60 and 52%, respectively). The more oxidizing environment in the rhizosphere might have played an important role, but other rhizosphere effects, beside rhizosphere aeration, appeared to be important also. Carbamazepine and declofenac showed low removal efficiencies in CWs, and this is attributable to their higher hydrophobicity. The fact that the removal of these compounds could be explained by the sorption onto the available organic surfaces, explains why there was no significant difference ($p > 0.05$) in their removal efficiencies between planted as compared to unplanted beds.

Two pilot scale horizontal subsurface flow constructed wetlands (HSSF-CWs) near Lecce, Italy, planted with different macrophytes (*P. australis* and *T. latifolia*) and an unplanted control were assessed for their effectiveness in removing paracetamol. The *P* bed exhibited a range of paracetamol removals from 51.7% for a hydraulic loading rate (HLR) of 240 mm/d to 87% with 120 mm/d HLR and 99.9% with 30 mm/d. The *Typha* bed showed a similar behavior with percentages of removal slightly lower, ranging from 46.7% (HLR of 240 mm/d) to >99.9% (hydraulic loading rate of 30 mm/d). At the same HLR values the unplanted bed removed between 51.3 and 97.6% of the paracetamol [139].

Hijosa-Valser et al. [88] run seven mesocosm-scale CWs, differing in their design characteristics to assess their efficiency to remove antibiotics from urban raw wastewater. All the studied types of CWs were efficient for the removal of sulfamethoxazole (59 ± 30–87 ± 41%), as found in the WWTP, and, in addition, they removed trimethoprim (65 ± 21–96 ± 29%). The elimination of other antibiotics in CWs was limited by the specific system-configuration: amoxicillin (45 ± 15%) was only eliminated by a free-water (FW) subsurface flow (SSF) CW planted with *T. angustifolia*; doxycycline was removed in FW systems planted with *T. angustifolia* (65 ± 34–75 ± 40%), in a *P. australis*-floating macrophytes system (62 ± 31%) and in conventional horizontal SSF-systems (71 ± 39%); clarithromycin was partially eliminated by an unplanted FW-SSF system (50 ± 18%); erythromycin could only be removed by a *P. australis*-horizontal SSF system (64 ± 30%); and ampicillin was eliminated by a *T. angustifolia-floating* macrophytes system (29 ± 4%). Lincomycin was not removed by any of the systems (WWTP or CWs).

A surface flow constructed wetland in Can Cabanyes (Granollers, Catalonia, northeastern Spain) was created as a part of a series of activities aimed at restoring a highly impacted fluvial peri-urban zone [31]. The system was fed with a small part of the secondary effluent, which was not completely nitrified, from an urban wastewater treatment plant. The results for PPCPs demonstrated that the wetland has a good capacity for removing a large variety of these compounds; the removal efficiencies were higher than 70% for most of them, with the exception of clofibric acid (34%) and carbamazepine (39%). These results were in agreement with the studies of Matamoros et al. [6] carried out at the same initial concentration as shown in Table 3.6. The effect of ciprofloxacin on the development, function and stability of bacterial communities in four mesocosm wetlands planted with *P. australis* was investigated [140]. The results showed that ciprofloxacin exposure may have an adverse effect on the inherent bacterial communities in wetland systems initially reducing their ability to assimilate anthropogenic carbon-based compounds; however, normal functionality may resume after a 2–5 week period.

## 3.4 Concluding Remarks

Constructed wetlands technology for removal of trace pollutants is an emerging field. However as shown in this chapter, recent studies proved that CWs have a high capacity for removing a large range of these compounds, some of them taken in consideration for their regulation by the European Commission (i.e. galaxolide, diclofenac and carbamazepine).

The mechanisms governing the removal of organic chemicals are not well established yet. There is no agreement on physical, chemical and biological processes that attend to attenuation in constructed wetlands. Their knowledge is crucial for further optimization of system designs and operational models. Therefore much research needs to be addressed in this field.

**Acknowledgments** The authors (G. Y. Töre and S. Meriç) would thank Namik Kemal University Reserch Fund (Project no: NKUBAP.00.17.AR.11.02) for supporting this work.

# References

1. Ávila C, Pedescoll A, Matamoros V, Bayona JM, García J (2010) Capacity of a horizontal subsurface flow constructed wetland system for the removal of emerging pollutants: an injection experiment. Chemosphere 81(9):1137–1142
2. Agudelo RM, Penuela G, Aguirre NJ, Moratod J, Jaramill ML (2010) Simultaneous removal of chlorpyrifos and dissolved organic carbon using horizontal sub-surface flow pilot wetlands. Ecol Eng 36:1401–1408
3. Song HL, Nakano K, Taniguki T, Nomura M, Nishimura O (2009) Estrogenal removal from treated municpal effluent in small scale constructed wetlands with different depth. Bioresour Technol 100:2945–2951
4. Fatta-Kassinos D, Meriç S, Nikolaou A (2011) Pharmaceutical residues in environmental waters and wastewater: current state of knowledge and future research. Anal Bioanal Chem. doi:10.1007/s00216-010-4300-9
5. Matamoros V, Caselles-Osorio A, Garcia J, Bayona JM (2008) Behaviour of pharmaceutical products and biodegradation intermediates in horizontal subsurface flow constructed wetland. A microcosm experiment. Sci Total Environ 394:171–176
6. Matamoros V, Garcia J, Bayona JM (2008) Organic micropollutant removal in a full scale surface flow constructed wetland fed with secondary effluents. Water Res 42:653–660
7. Hijosa-Valsero M, Matamoros V, Sidrach-Cardona R, Martín-Villacorta J, Bécares E, Bayona JM (2010) Comprehensive assessment of the design configuration of constructed wetland for the removal of pharmaceuticals and personal care products from urban wastewater. Water Res 44:3669–3678
8. De Feo G, Lofrano G, Belgiorno V (2005) Treatment of high strength wastewater with vertical flow constructed wetland filters. Water Sci Technol 51(10):139–147
9. Imfeld G, Braeckevelt M, Kuschk P, Richnow HH (2009) Monitoring and assessing processes of organic chemicals removal in constructed Wetlands. Chemosphere 74:349–362
10. Brix H (1997) Do machrophytes play a role in constructed treatment wetlands? Water Sci Technol 35(5):11–17
11. Newman LA, Reynolds CM (2004) Phytodegradation of organic compounds. Curr Opin Biotechnol 15:225–230
12. Susarla S, Medina VF, McCutcheon SC (2002) Phytoremediation: an ecological solution to organic chemical contamination. Ecol Eng 18:647–658
13. Bankston JL, Sola DL, Komor AT, Dwyer DF (2002) Degradation of trichloroethylene in wetland microcosms containing broad-leaved cattail and eastern cottonwood. Water Res 36:1539–1546
14. Wang X, Dossett MP, Gordon MP, Strand SE (2004) Fate of carbon tetrachloride during phytoremediation with poplar under controlled field conditions. Environ Sci Technol 38:5744–5749
15. Brix H, Schierup HH (1989) The use of macrophytes in water pollution control. Ambio 18:100–107
16. Vymazal J, Kröpfelová L (2008) Wastewater treatment in constructed wetlands with horizontal sub-surface flow; Springer: Dordrecht
17. Seidel K (1955) Die Flechtbinse scirpus lacustris. In: Ökologie, Morphologie und Entwicklung, ihre Stellung bei den Volkern und ihre wirtschaftliche Bedeutung; Schweizerbart'scheVerlagsbuchnadlung: Stuttgart, Germany, pp 37–52
18. Seidel K (1966) Reinigung von Gewässern durch höhere Pflanzen. Naturwissenschaften 53:289–297
19. Seidel K (1976) Macrophytes and water purification. In: Tourbier J, Pierson RW (eds) Biological control of water pollution. Pennsylvania University Press, Philadelphia, pp 109–122
20. Seidel K (1961) Zur Problematik der Keim-und Pflanzengewasser. Verh Internat Verein Limnol 14:1035–1039

21. De Jong J (1976) The Purification of Wastewater with the Aid of Rush or Reed Ponds. In: Tourbier J, Pierson RW (eds) Biological control of water pollution. Pennsylvania University Press, Philadelphia, pp 133–139
22. Kadlec RH, Wallace SD (2008) Treatment wetlands, 2nd edn. CRC Press, Boca Raton
23. Mander U, Jenssen P (2003) Constructed wetlands for wastewater treatment in cold climates. WIT Press, Southampton
24. Alvarez JA, Ruiz I, Soto M (2008) Anaerobic digesters as a pretreatment for constructed wetlands. Ecol Eng 33(1):54–67
25. Higgins JM, El-Qousey D, Abul-Azm AG, Abdelghaffar M (2001) Lake manzala engineered wetland, Egypt in proceedings of the 2001. Wetlands engineering & river restoration conference, 27–31 Aug, Reno, Nevada
26. Hoffmann H, Platzer C, Heppeler D, Barjenbrunch M, Tränckner J, Belli P (2002) Combination of anaerobic treatment and nutrient removal of wastewater in Brazil. In: Proceedings of the 3rd water world congress. Melbourne, Australia, pp 9–12
27. Sengorur B, Ozdemir S (2006) Performance of a constructed wetland system for the treatment of domestic wastewater. Fresenius Environ Bull 15(3):242–244
28. VonSperling M (1996) Comparison among the most frequently used systems for wastewater treatment in developing countries. Water Sci Technol 33(3):59–72
29. Zhang XL, Zhang S, He F, Cheng SP, Liang W, Wu ZB (2007) Differentiate performance of eight filter media in vertical flow constructed wetland: removal of organic matter, nitrogen and phosphorus. Fresenius Environ Bull 16(11B):1468–1473
30. Greenway M (2005) The role of constructed wetlands in secondary effluent treatment and water reuse in subtropical and arid Australia. Ecol Eng 25:501–509
31. Llorens E, Matamoros V, Domingo V, Bayona JM, García J (2009) Water quality improvement in a full-scale tertiary constructed wetland: effects on conventional and specific organic contaminants. Sci Total Environ 407(8):2517–2524
32. Karathanasis AD, Johnson CM (2003) Metal removal potential by three aquatic plants in an acid mine drainage wetland. Mine Water Environ 22:22–30
33. Juwarkar AS, Oke B, Juwarkar A, Patnaik SM (1995) Domestic wastewater treatment through constructed wetlands in India'. Water Sci Technol 32(3):291–294
34. Blazejewski R, Murat Blazejewska S (1997) Soil clogging phenomena in constructed wetlands with subsurface flow. Water Sci Technol 35(5):183–188
35. Caselles-Osorio A, Garcia J (2007) Effect of physico-chemical pretreatment on the removal efficiency of horizontal subsurface-flow constructed wetlands. Environ Pollut 146(1):55–63
36. Dahab MF, Surampalli RY (2001) Subsurface-flow constructed wetlands treatment in the plains: five years of experience. Water Sci Technol 44(11–12):375–380
37. Garcia J, Rousseau D, Caselles-Osorio A, Story A, Pauw ND, Vanrolleghem P (2007) Impact of prior physico-chemical treatment on the clogging process of subsurface flow constructed wetlands: model-based evaluation. Water Air Soil Pollut 185(1–4):101–109
38. Hua GF, Zhu W, Zhao LF, Huang JY (2010) Clogging pattern in vertical-flow constructed wetlands: insight from a laboratory study. J Hazard Mater 180(1–3):668–674
39. Langergraber G, Haberl R, Laber J, Pressl A (2003) Evaluation of substrate clogging processes in vertical flow constructed wetlands. Water Sci Technol 48(5):25–34
40. Nguyen LM (2000) Organic matter composition, microbial biomass and microbial activity in gravel-bed constructed wetlands treating farm dairy wastewaters. Ecol Eng 16(2):199–221
41. Winter KJ, Goetz D (2003) The impact of sewage composition on the soil clogging phenomena of vertical flow constructed wetlands. Water Sci Technol 48(5):9–14
42. Kayser K, Kunst S (2005) Processes in vertical-flow reed beds: nitrification, oxygen transfer and soil clogging. Water Sci Technol 51(9):177–184
43. Batchelor A, Loots P (1997) A critical evaluation of a pilot scale subsurface flow wetland: 10 years after commissioning. Water Sci Technol 35(5):337–343
44. Vymazal J (2005) Horizontal sub-surface flow and hybrid constructed wetlands for wastewater treatment. Ecol Eng 25:478–490

45. Vymazal J (2005) Constructed wetlands for wastewater treatment. Ecol Eng 25(5): 475–477

46. Ge Y, Li SP, Niu XY, Yue CL, Xu QS, Chang J (2007) Sustainable growth and nutrient uptake of plants in a subtropical constructed wetland in southeast China. Fresenius Environ Bull 16(9A):1023–1029

47. Neralla S, Weaver RW, Lesikar BJ, Persyn RA (2000) Improvement of domestic wastewater quality by subsurface flow constructed wetlands. Bioresour Technol 75:19–25

48. Vymazal J (2002) The use of sub-surface constructed wetlands for wastewater treatment in the Czech Republic: 10 years experience. Ecol Eng 18:633–646

49. Yue CL, Chang J, Ge Y, Zhu YM (2004) Treatment efficiency of domestic wastewater by vertical/reverse-vertical flow constructed wetlands. Fresenius Environ Bull 13(6):505–507

50. Amuda OS, Amoo IA, Ajay OO (2005) Performance optimization of coagulant/flocculent in the treatment of wastewater from a Beverage industry. J Hazard Mater 129(1–3):69–72

51. Meers E, Rousseau D, Lesage E, Demeersseman E, Tack F (2006) Physicochemical P removal from the liquid fraction of pig manure as an intermediary step in manure processing. Water Air Soil Pollut 169(1–4):317–330

52. Selcuk H, Kaptan D, Meric S (2004) Coagulation of textile finishing industry wastewater using alum and Fe(III) salts. Fresenius Environ Bull 13:1045–1048

53. Gurses A, Yalcin M, Dogar C (2003) Investigation on settling velocities of aluminium hydroxide-dye flocs. Fresenius Environ Bull 12:16–23

54. Sarparastzadeh H, Saeedi M, Naeimpoor F, Aminzadeh B (2007) Pretreatment of municipal wastewater by enhanced chemical coagulation. Int J Environ Res 1(2):104–113

55. Meric S, Guida M, Mattei ML, Anselmo A, Melluso G (2002) Evaluation of coagulation/ flocculation process in S. Giovanni a Teduccio municipal wastewater treatment plant. Fresenius Environ Bull 11:906–909

56. Mungasavalli DP, Viraraghavan T (2006) Constructed wetlands for stormwater management: a review. Fresenius Environ Bull 15(11):1363–1372

57. Revitt DM, Shutes RBE, Jones RH, Forshaw M, Winter B (2004) The performance of vegetative treatment systems for highway runoff during dry and wet conditions. Sci Tot Environ 334–335:261–270

58. Boutin C (1987) Domestic wastewater treatment in tanks planted with rooted macrophytes: case study, description of the system, design criteria, and efficiency. Wat Sci Tech 19:29–40

59. Bulc TG (2006) Long term performance of a constructed wetland for landfill leachate treatment. Ecol Eng 26:365–374

60. Kinsley CB, Crolla AM, Kuyucak N, Zimmer M, Lafleche A (2006) Nitrogen dynamics in a constructed wetland system treating landfill leachate. In: Proceedings of the 10th international conference on wetland systems for water pollution control; MAOTDR: Lisbon, Portugal, pp 295–305

61. Reeb G, Werckmann M (2005) First performance data on the use of two pilot-constructed wetlands for highly loaded non-domestic sewage. In: Vymazal J (ed) Natural and constructed wetlands: nutrients, metals and management. Backhuys Publishers, Leiden, pp 43–51

62. Sôroko M (2005) Treatment of wastewater from small slaughterhouse in hybrid constructed wetlands system. In: Toczyłowska I, Guzowska G (eds) Proceedings of the workshop wastewater treatment in wetlands. Theoretical and practical aspects, Gdańsk University of Technology Printing Office: Gdansk, p 171–176

63. Lin YF, Jing SR, Lee DY, Wang TW (2002) Nutrient removal from aquaculture wastewater using a constructed wetlands system. Aquaculture 209:169–184

64. Lin YF, Jing SR, Lee DY (2003) The potential use of constructed wetlands in a recirculating aquaculture system for shrimp culture. Environ Poll 123:107–113

65. Masi F, Conte G, Martinuzzi N, Pucci B (2002) Winery high organic content wastewaters treated by constructed wetlands in mediterranean climate. In: Proceedings of the 8th international conference on wetland systems for water pollution control, University of Dar-es-Salaam: Dar-es-Salaam, Tanzania, pp 274–282

66. Veenstra S (1998) The Netherlands. In: Vymazal, J, Brix H, Cooper PF, Green MB, Haberl R (eds) Constructed wetlands for wastewater treatment in Europe. Backhuys Publishers, Leiden

67. Ewel KC, Odum HT (1984) Cypress swamps. University of Florida Press, Gainesville

68. Kadlec RH, Tilton DL (1979) The use of freshwater wetlands as a tertiary wastewater treatment alternative. CRC Crit Rev Env Control 9:185–212

69. Odum HT, Ewel KC, Mitsch WJ, Ordway JW (1977) Recycling treated sewage through cypress wetlands in Florida. In: D'Itri FM (ed) Wastewater renovation and reuse. Marcel Dekker, New York pp 35–67

70. Lesavre J, Iwema A (2002) Dewatering of sludge coming from domestic wastewater treatment plant by planted sludge beds. French situation. In: Proceedings of the 8th international conference on wetland systems for water pollution control, Arusha, Tanzania, pp 1193–1205

71. Liénard A, Duchène P, Gorini D (1995) A study of activated sludge dewatering in experimental Reed-planted or unplanted sludge drying beds. Wat Sci Tech 32:251–261

72. Molle P, Lienard A, Boutin C, Merlin G, Iwema A (2005) How to treat raw sewage with CW: an overview of the French Systems. Wat Sci Tech 51(9):11–22

73. Paing J, Voisin J (2005) Vertical flow constructed wetlands for municipal wastewater and septage treatment in French rural area. Wat Sci Tech 51(9):145–157

74. Garcia J, Morato J, Bayona JM, Aguirre P (2004) Performance of horizontal sub-surface flow CW with different depth. In: Proceedings of the 9th international conference on wetland systems for water pollution control, Avignon, France, pp 269–278

75. Matos J, Santos S, Dias S (2002) Small wastewater systems in Portugal: challenges, strategies and trends for the future" in Proceedings of "Small wastewater technologies and management for the Mediterranean area, Seville, Spain

76. Mandi L (1996) The use of aquatic macrophytes in the treatment of wastewater under arid climate: marrakech experiment In: Proceedings of the 5th international conference on wetland systems for water pollution control, Wien, Austria

77. Conte G, Martinuzzi N, Giovannelli L, Pucci B, Masi F (2001) Constructed wetlands for wastewater treatment in central Italy. Water Sci Technol 44(11–12):339–343

78. Masi F, Martinuzzi N, Loiselle S, Peruzzi P, Bacci M (1999) The tertiary treatment pilot plant of Publishser (Florence–Tuscany): a multistage experience". Water Sci Technol 40(3): 195–202

79. Masi F, Bendoricchio G, Conte G, Garuti G, Innocenti A, Franco D, Pietrelli L, Pineschi G, Pucci Romagnolli BF (2000) Constructed wetlands for wastewater treatment in Italy: state-of-the-art and obtained results, Proceedings of the 7th IWA international conference on wetland systems for water pollution control, Orlando, pp 979–985

80. Masi F, Conte G, Lepri L, Martellini T, Del Bubba M (2004) Endocrine disrupting chemicals (EDCs) and Pathogens removal in an Hybrid CW system for a tourist facility wastewater treatment and reuse". In: Proceedings of the 9th IWA international conference on wetland systems for water pollution control. Avignon (France) 2:461–468

81. Awad AM, Saleh HI (2001) Evaluating contaminants removal rates in sub-surface flow constructed wetland in egypt". In: Proceedings of the 2001 wetlands engineering and river restoration conference, Aug 27–31, Reno, Nevada

82. Brenner A, Messalem R (2002) Wastewater treatment and reuse in Israel: policy and applications. In: Proceedings of "small wastewater technologies and management for the mediterranean area", Seville, Spain

83. Bulc TG (2002) Development of CW in Slovenia. In: proceedings of "small wastewater technologies and management for the mediterranean area", Seville, Spain

84. Bulc TG, Zupancic M, Vrhovsek D (2003) CW experiences in Slovenia: development and application. In: Proceedings of the conference constructed wetlands: applications and future possibilities, Volterra, Italy, pp 90–105

85. Shalabi M (2004) CW in croatian adriatic area In: Proceedings of 9th international conference on wetland systems for water pollution control, Avignon, France, pp 307–314

86. Papadopoulos A (2002) The Nagref experimental station in the thessaloniki greece". In: Proceedings of "Small wastewater technologies and management for the Mediterranean area", Seville, Spain
87. Yildiz C, Korkusuz AE, Arikan Y, Demirer GN (2004) CW for municipal wastewater treatment: a study from Turkey. In: Proceedings of 9th international conference on wetland systems for water pollution control, Avignon, France, 26–30 Sept 2004, pp 193–202
88. Hijosa-Valsero M, Matamoros V, Sidrach-Cardona R, Martín-Villacorta J, Bécares E, Bayona JM (2011) Removal of antibiotics from urban wastewater by constructed wetland optimization. Chemosphere 83(2011):713–719
89. Brix H (1995) Function of macrophytes in constructed wetlands. Water Sci Technol 29(4): 71–78
90. Cooper P (1999) A review of the design and performance of vertical-flow and hybrid reed bed treatment systems'. Water Sci Technol 40(3):1–9
91. EPA (2004) Guidelines for water reuse. U.S. EPA 625/R-04/108
92. Luedertiz V, Eckert E, Lange-Weber M, Lange A, Gerseberg R (2001) Nutrient removal efficiency and resource economics of vertical flow and horizontal flow constructed wetlands. Ecol Eng 18:157–171
93. Mandi L, Bouhoum K, Ouazzani N (1998) Application of constructed wetlands for domestic wastewater treatment in an arid climate. Water Sci Technol 38(1):379–387
94. World Health Organization (WHO) (1989) Health guidelines for the use of wastewater and aquaculture: report of a WHO scientific group, WHO Technical Report Series 778—Health Organization. Switzerland, Geneva
95. World Health Organization (WHO) (2001) Water Quality: guidelines. IWA publishing, Standards and Health, London
96. Vymazal J (2010) Constructed wetlands for wastewater treatment: five decades of experience. Environ Sci Technol 45:61–69
97. Wallace SD (2002) On-site Remediation of Petroleum Contact Wastes Using Subsurface-flow Wetlands. In: Nehring KW, Brauning SE (eds) Wetlands and Remediation II. Battelle Press: Columbus, OH, USA, pp 125–132
98. Ji G, Sun T, Zhou Q, Sui X, Chang S, Li P (2002) Constructed subsurface slow wetland for treating heavy oil-produced water of the Liaohe Oilfield in China. Ecol Eng 18:459–465
99. Sands Z, Gill LS, Rust R (2000) Effluent Treatment Reed Beds: results after Ten Years of Operation. In: Means JF, Hinchee RE (eds) Wetlands and Remediation. Battelle Press: Columbus, OH, USA
100. Thut RN (1993) Feasibility of Treating Pulp Mill Effluent with a Constructed Wetland. In: Moshiri GA (ed) Constructed Wetlands for Water Quality Improvement. Lewis Publishers: Boca Raton, FL, USA, pp 441–447
101. Lavigne RL, Jankiewicz J (2000) Artificial Wetland Treatment Technology and it's use in the Amazon River Forests of Ecuador. In: Proceedings of the 7th International Conference on Wetland Systems for Water Pollution Control; University of Florida: Gainesville, FL, USA, pp 813–820
102. Poggi-Varaldo HM, Gutiérez-Saravia A, Fernández-Villagómez G, Martínez-Pereda P, Rinderknecht-Seijas N (2002) A full-scale System with Wetlands for Slaughterhouse Wastewater Treatment. In: Nehring KW, Brauning SE (eds) Wetlands and Remediation II. Battelle Press: Columbus, OH, USA, pp 213–223
103. Davies TH, Cottingham PD (1992) The use of constructed wetlands for treating industrial effluent. In: Proceedings of the 3rd International Conference on Wetland Systems in Water Pollution Control; IAWQ and Australian Water and Wastewater Association: Sydney, Australia, pp 53.1–53.5
104. Calheiros CSC, Rangel AOSS, Castro PKL (2007) Constructed wetland systems vegetated with different plants applied to the treatment of tannery wastewater. Water Res 41: 1790–1798
105. Vrhovšek D, Kukanja V, Bulc T (1996) Constructed wetland (CW) for industrial waste water treatment. Water Res 30:2287–2292

106. Mantovi P, Marmiroli M, Maestri E, Tagliavini S, Piccinini S, Marmiroli N (2003) Application of a horizontal subsurface flow constructed wetland on treatment of dairy parlor wastewater. Bioresour Technol 88:85–94

107. Billore SK, Singh N, Ram HK, Sharma JK, Singh VP, Nelson RM, Das P (2001) Treatment of a molasses based distillery effluent in a constructed wetland in central India. Water Sci Tech 44:441–448

108. Finlayson M, Chick A, von Oertzen I, Mitchell D (1987) Treatment of piggery effluent by an aquatic plant filter. Biol Wastes 19:179–196

109. Strusevičius Z, Strusevičiene SM (2003) Investigations of wastewater produced on cattle-breeding farms and its treatment in constructed wetlands. In: Proceedings of the International Conference on Constructed and Riverine Wetlands for Optimal Control of Wastewater at Catchment Scale; Mander Ü, Vohla C, Poom A, (eds) University of Tartu, Institute of Geography: Tartu, Estonia, pp 317–324

110. Comeau Y, Brisson J, Réville J-P, Forget C, Drizo A (2001) Phosphorus removal from trout farm effluents by constructed wetlands. Wat Sci Tech 44:55–60

111. Schulz C, Gelbrecht J, Rennert B (2003) Treatment of rainbow trout farm effluents in constructed wetland with emergent plants and subsurface horizontal water flow. Aquaculture 217:207–221

112. Perdomo S, Bangueses C, Fuentes J, Castro J, Acevedo H, Michelotti C (2000) Constructed wetlands: a more suitable alternative for wastewater purification in Uruguayn Dairy Processing Industry. In: Proceedings of the 7th International Conference on Wetland Systems for Water Pollution Control; Reddy KR, Kadlec RH (eds) University of Florida and IWA: Gainesville, FL, USA, pp 1407–1415

113. Kern J, Brettar I (2002) Nitrogen Turnover in a Subsurface Constructed Wetland Receiving dairy Farm Wastewater. In: Pries J (ed) Treatment Wetlands for Water Quality Improvement. CH2M Hill Canada Limited: Waterloo, Canada, pp 15–21

114. Drizo A, Twohig E, Weber D, Bird S, Ross D (2006) Constructed wetlands for dairy effluent treatment in vermont: two years of operation. In: Proceedings of the 10th International Conference on Wetland Systems for Water Pollution Control; MAOTDR 2006: Lisbon, Portugal, pp 1611–1621

115. DuBowy P, Reaves P (1994) Constructed wetlands for animal waste management; Conservation Technology Information Center, U.S. Dept of Agriculture Soil Conservation Service, U.S. EPA Region V and Purdue University Agric Res Program: Lafayette, IN, USA

116. Knight RL, Payne VWE Jr, Borer RE, Clarke RA Jr, Pries JH (2000) (a). Constructed wetlands for livestock wastewater management. Ecol Eng 15:41–55

117. Knight RL, Clarke RA Jr, Bastian RK (2000) (b). Surface flow (sf) treatment wetlands as a habitat for wildlife and humans. Wat Sci Tech 44:27–38

118. Tanner CC, Nguyen ML, Sukias JPS (2005) Nutrient removal by a constructed wetland treating subsurface drainage from a grazed dairy pasture. Agric Ecosyst Environ 105: 145–162

119. Ramírez Masferrer JA (2002) Passive treatment of acid mine drainage at the La Extranjera Mine(Puertollano, Spain). Mine Water Environ 21:111–113

120. Sobolewski A (1996) Metal species indicate the potential of constructed wetlands for long-term treatment of mine drainage. J Ecol

121. Kiessig G, Küchler A, Kalin M (2003) Passive Treatment of Contaminated Water from Uranium Mining and Milling. In: Mander Ü, Vohla C, Poom A (eds) Constructed and Riverine Wetlands for Optimal Control of Wastewater at Catchment Scale. University of Tartu: Tartu, Estonia, p 116

122. ÓSullivan AD, Moran BM, Otte ML (2004) Accumulation and fate of contaminants (Zn, Pb, Fe and S) in substrates of wetlands constructed for treating mine wastewater. Water Air Soil Pollut 157:345–364

123. Litchfield DK, Schatz DD (1989) Constructed Wetlands for Wastewater Treatment at Amoco Oil Company's Mandan, North Dakota refinery. In: Hammer DA (ed) Constructed Wetlands for Wastewater Treatment. Lewis Publishers: Chelsea, MI, USA, pp 233–237

124. Lakatos G (1998) Hungary. In Constructed Wetlands for Wastewater Treatment in Europe; Vymazal J, Brix H, Cooper PF, Green MB, Haberl R (eds) Backhuys Publishers: Leiden, The Netherlands, pp 191–206
125. Tettleton RP, Howell FG, Reaves RP (1993) Performance of a Constructed Marsh in the Tertiary Treatment of Bleach Kraft Pulp Mill Effluent: results of a 2-year Pilot Project. In: Moshiri GA (ed) Constructed Wetlands for Water Quality Improvement. CRC Press/ Lewis Publishers: Boca Raton, FL, USA, pp 437–440
126. Tonderski KS, Grönlund E, Billgren C (2005) Management of sugar effluent in the Lake Victoria Region. In: Proceedings of the Workshop Wastewater treatment in Wetlands. Theoretical and Practical Aspects; Toczyłowska I, Guzowska G, (eds) Gdańsk University of Technology Printing Office: Gdansk, Poland, pp 177–184 ,
127. Kapellakis IE, Tsagarakis KP, Angelakis AN (2004) Performance of free water surface constructed wetlands for Olive Mill Wastewater Treatment. In: Proceedings of the 9th International Conference on wetland systems for water pollution control; ASTEE: Lyon, France, pp 113–120
128. Maine MA, Sune N, Hadad H, Sánchez G, Bonetto C (2006) Nutrient and metal removal in a constructed wetland for waste-water treatment from a metallurgic industry. Ecol Eng 26:341–347
129. Maine MA, Sune N, Hadad H, Sánchez G, Bonetto C (2007) Removal efficiency of a constructed wetland for wastewater treatment according to vegetation dominance. Chemosphere 68: 1105–1113
130. Fontoulakis MS, Terzakis S, Kalogerakis N, Manios T (2009) Removal of polyciclic aromatic hydrocarbons and linear alkylbenzene sulfonates from domestic wastewater in pilot constructed wetlands and a gravel filter. Ecol Eng 35:1702–1709
131. Masi F, Conte G, Lepri L, Martellini T, Del Bubba M (2005) Endocrine disrupting chemicals (EDCs) and pathogens removal in an hybrid CW system for a tourist facility in wastewater treatment and reuse, http://www.iridra.it/.
132. Huang Y, Latorre A, Barcelo D, Garcia J, Aguirre P, Mujeriego R, Bayona JM (2004) Factors affecting linear alkylbenzene sulfonates removal in subsurface flow constructed wetlands. Environ Sci Technol 38:2657–2663
133. Kantawanichkul S, Wara-Aswapati S (2005) LAS removal by a horizontal flow constructed wetland in tropical climate. In: Vymazal J (ed) Natural and constructed wetlands: nutrients, metals and management. Backhuys Publishers: Leiden
134. Matamoros V, Bayona JM (2006) Elimination of pharmaceuticals and personal careproducts in subsurface flow constructed wetlands. Environ Sci Technol 40(18):5811–5816
135. Zhang DQ, Tan SK, Gersberg RM, Sadreddini S, Zhu J, Tuan NA (2011) Removal of pharmaceutical compounds in tropical constructed wetlands. Ecol Eng 37(3):460–464
136. Conkle JL, White JR, Metcalfe CD (2008) Reduction of pharmaceutically active compounds by a lagoon wetland wastewater treatment system in Southeast Louisiana. Chemosphere 73:1741–1748
137. Dordio AV, Estevao Candeias AJ, Pinto AP, Teixeira da Costa C, Palace Carvalho AJ (2009) Preliminary media screening for application in the removal of clofibric acid, carbamazepine and ibuprofen by SSF-constructed wetlands. Ecol Eng 35:290–302
138. Dordio AV, Duarte C, Barreiros M, Palace Carvalho AJ, Pinto AP, Teixeira da Costa C (2009) Toxicity and removal efficiency of pharmaceutical metabolite clofibric acid by Typha spp—potential use for phytoremediation? Bioresour Technol 100:1156–1161
139. Ranieri E, Verlicchi P, Young TM (2011) Paracetamol removal in subsurface flow constructed wetlands. J Hydrol 404:130–135
140. Weber KP, Mitzel MR, Slawson RM, Legge RL (2011) Effect of ciprofloxacin on microbiological development in wetland mesocosms. Water Res 45:3185–3196

# Chapter 4
# Removal of Pesticides from Water and Wastewater by Solar-Driven Photocatalysis

Sixto Malato, Manuel I. Maldonado, Isabel Oller and Ana Zapata

**Abstract** This chapter deals with the use of sunlight to produce •OH radicals by photocatalysis and its application to the removal of pesticides from water. The systems necessary for performing solar photocatalysis based on compound parabolic collectors are described and it outlines the basic components of these plants. It reports a step-by-step research methodology describing the analytical tools to infer the reaction mechanisms, pathway and kinetics and the application of various techniques for determining biodegradability and toxicity. Besides, it underlines the importance of: (i) using acute toxicity bioassays, for stating biocompatibility of the treated water with the environment and (ii) using photocatalysis as a pre-treatment step, if the intermediates resulting from the reaction are readily degraded by microorganisms (biotreatment).

**Keywords** Advanced oxidation processes · Compound parabolic concentrators · Photo-Fenton · Titanium dioxide

## 4.1 Introduction

Advanced Oxidation Processes (AOPs) may be used for decontamination of water containing organic pollutants, classified as biorecalcitrant. These methods rely on the formation of highly reactive chemical species which degrade even the most recalcitrant molecules into biodegradable compounds. Although there are different

S. Malato (✉) · M. I. Maldonado · I. Oller · A. Zapata
Plataforma Solar de Almería (CIEMAT), Carretera Senés, km 4,
04200, Tabernas, Almería, Spain
e-mail: Sixto.malato@psa.es

G. Lofrano (ed.), *Emerging Compounds Removal from Wastewater*,
SpringerBriefs in Green Chemistry for Sustainability,
DOI: 10.1007/978-94-007-3916-1_4, © Malato, Maldonado, Oller, Zapata 2012

reacting systems [1], all of them are characterised by the same chemical feature: production of hydroxyl radicals ($^{\bullet}$OH), which are able to oxidise and mineralise almost any organic molecule, yielding $CO_2$ and inorganic ions. Rate constants ($k_{OH}$, $r = k_{OH}$ [$^{\bullet}$OH] C) for most reactions involving hydroxyl radicals in aqueous solution are usually on the order of $10^6$–$10^9$ $M^{-1}$ $s^{-1}$. They are also characterised by their not-selective attack, which is a useful attribute for wastewater treatment and solution of pollution problems. The versatility of the AOPs is also enhanced by the fact there are different ways of producing hydroxyl radicals, facilitating compliance with the specific treatment requirements. Methods based on UV, $H_2O_2$ and $O_3$ combinations use photolysis of $H_2O_2$ and ozone to produce the hydroxyl radicals. Other methods, like heterogeneous photocatalysis and homogeneous photo-Fenton, are based on the use of a wide-band-gap semiconductor and addition of $H_2O_2$ to dissolved iron salts, respectively, and irradiation with UV–VIS light [2]. Both processes are of special interest since sunlight can be used for them.

The use of pesticides has risen dramatically, with the production nearly doubled every 5 years since 1975. UN reports estimates that of all pesticides used in agriculture, less than 1% actually reaches the crops. This results in the uncontrolled disposal of used products that will produce contaminated soils and waters close to the contaminant source. Their persistence in natural waters [3] has led to a search for a method to degrade them into environmentally compatible compounds. Unlike the low-level contamination involved in drinking water, wastewater from agricultural or industrial activities may be highly contaminated. The major sources of pollution are wastewater from agricultural industries, pesticides formulating and manufacturing plants. Wastewater from those sources may contain pesticides at levels as high as several hundred of mg/L. The main characteristics of them are extreme toxicity, low volume and well-defined location. Such sources may be ideally treated in small-scale treatment units. As consequence, low cost and at hand technologies are strongly urged to be developed to on site treatment. AOPs, are well known for their capacity to oxidize and mineralise pesticides [4]. As the process costs may be considered the main obstacle to their commercial application, several promising cost-cutting approaches have been proposed, such as integration of AOPs as part of a treatment train. In the typical basic process design approach an AOP pretreats pesticide wastewater, and once biodegradability has been achieved, the effluent is transferred to a cheaper biological treatment. Other proposed cost-cutting measures are the use of renewable energy sources, i.e., sunlight as the irradiation source for running the AOP.

The publications regarding the photocatalytic process rose continuously over the last years surpassing meanwhile a total number of more than 3,000 peer-reviewed publications per year. Though such a simple search does not necessarily include every single article correctly, it still serves to prove the general trend of an increasing interest of the scientific community. Figure 4.1 shows the evolution of these publication activities. Figure 4.1 also illustrates the literature that takes into account the possibility of driving the process with solar radiation. This fact is due to that a priori the photocatalytic processes ($TiO_2$ photocatalysis and photo-Fenton) seems to be the most apt of all AOPs to be driven by sunlight.

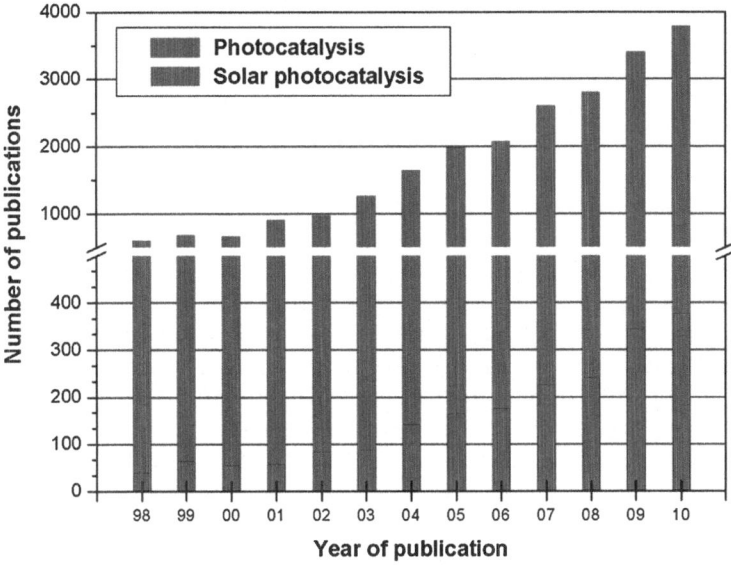

**Fig. 4.1** Publications treating photocatalysis and the share treating solar-driven photocatalysis (*source*: www.scopus.com, 2011. )

In this chapter we highlight some of the science and technology being developed to improve the solar photocatalytic decontamination of water containing pesticides.

## 4.2 Solar Photocatalysis Fundamentals

The heterogeneous solar photocatalytic detoxification process consists of making use of the near-ultraviolet (UV) band of the solar spectrum (wavelength shorter than 400 nm), to photo-excite a semiconductor catalyst in contact with water and in the presence of oxygen. Under these circumstances, oxidizing species, either bound $^{\bullet}OH$ or free holes, react with oxidizable contaminants. With a typical UV-flux near the surface of the earth of 20–30 $W/m^2$ the sun puts 0.2–0.3 mol photons $m^{-2} h^{-1}$ in the 300–400 nm range at the process disposal [5]. Although there are many different sources of $TiO_2$, Degussa (now Evonik) P25 $TiO_2$ has effectively become a standard [6] because it has (i) a reasonably well-defined nature (i.e. typically a 70:30 anatase:rutile mixture, non-porous, BET surface area $55 \pm 15$ $m^2/g$, average particle size 30 nm) and (ii) a substantially higher photocatalytic activity than most other readily available (commercial) $TiO_2$. Other semiconductor particles, e.g., CdS or GaP absorb larger fractions of the solar spectrum and can form chemically activated surface-bond intermediates, but unfortunately, these photocatalysts are degraded during the repeated catalytic cycles involved in heterogeneous photocatalysis generating toxic dissolved heavy metals in water.

For the treatment of industrial wastewater Fenton and Fenton-like processes are probably among the since longest and most applied advanced oxidation processes [7] and first proposals for wastewater treatment applications were reported in the sixties of the past century. Yet, it was not until the early nineties of the last century, when the discoveries of scientists working in the field of environmental sciences published results on the role of iron in atmospheric chemistry, which called the attention of scientists and engineers working in the wastewater treatment field. Soon afterwards, the first reports of the application of the photo-Fenton process (or photoassisted/light enhanced Fenton process) in wastewater treatment were published by the groups of Pignatello, Lipcznska-Kochany, Kiwi, Pulgarín and Bauer [7]. The primary step of the photoreduction of dissolved ferric iron is a ligand-to-metal charge-transfer reaction. Subsequently, intermediate complexes dissociate by means of irradiation forming $Fe^{+2}$. The ligand can be any Lewis base able to form a complex with ferric iron ($OH^-$, $H_2O$, $R–COO^-$, $R–OH$, $R–NH_2$ etc.). Depending on the reacting ligand, the product may be a hydroxyl radical such as in Eq. 4.1 or another radical derived from the ligand. The direct oxidation of an organic ligand is possible as shown for carboxylic acids in Eq. 4.2. Depending on the ligand the ferric iron complex has different light absorption properties and consequently, the pH plays a crucial role in the efficiency of the photo-Fenton reaction, because it strongly influences which complexes are formed. Thus, pH 2.8 was frequently postulated as an optimum pH for photo-Fenton treatment, because at this pH precipitation does not take place yet and the dominant iron species in solution is $[Fe(OH)]^{2+}$, the most photoactive ferric iron−water complex.

$$[Fe(OH)]^{2+} + hv \rightarrow Fe^{2+} + OH^{\bullet} \tag{4.1}$$

$$[Fe(OOC - R)]^{2+} + hv \rightarrow Fe^{2+} + CO_2 + R^{\bullet} \tag{4.2}$$

## 4.3 Solar Photocatalysis as Green Technology

Since 1990 there has been a clarification of the kind of solar technology, which should be involved, in solar AOPs [8]. The question was if it is necessary to concentrate the radiation for the photocatalysis technology and if a non-concentrating collector can be as efficient as concentrating ones. The reason of using one-sun systems for water treatment is firmly based on two factors, first the high percentage of UV photons in the diffuse component of solar radiation and second the low order dependence of rates on light intensity. For many of the solar detoxification system components, the equipment is identical to that used for other types of water treatment and construction materials are commercially available.

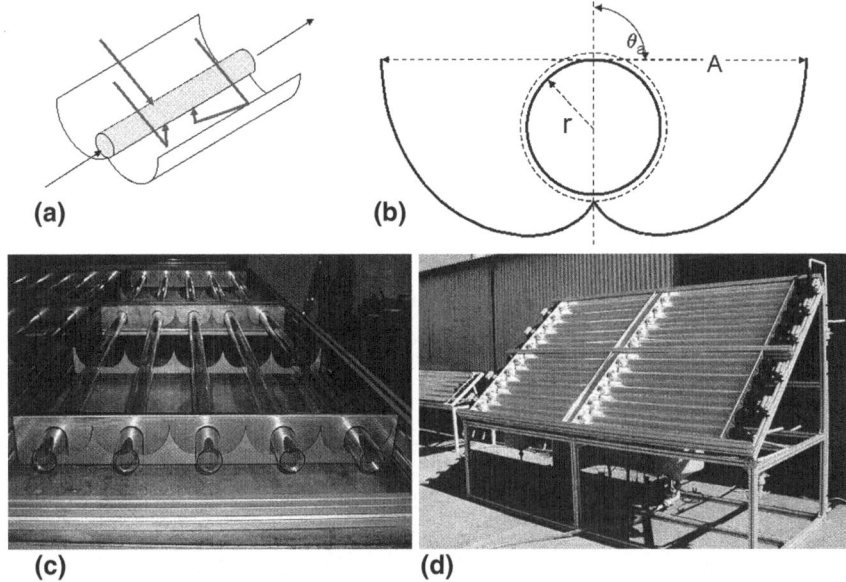

**Fig. 4.2** Design concepts for solar water photocatalytic reactors: **a** compound parabolic collector, **b** schematic drawing of CPC with a semi-angle of acceptance of 90°, **c**, **d** photographs of compound parabolic collector during fabrication and installed

Most piping may be made of polyethylene or polypropylene, avoiding the use of metallic or composite materials that could be degraded by the oxidant conditions of the process. Neither must materials be reactive, interfering with the photocatalytic process. All materials used must be inert to degradation by UV solar light. Photocatalytic reactors must transmit UV light efficiently because of the process requirements. The choice of materials that are both transmissive to UV light and resistant to its destructive effects is limited. Common materials that meet these requirements are fluoropolymers, acrylic polymers and several types of glass. Borosilicate glass has good transmissive properties in the solar range with a cut-off of about 285 nm. Therefore, such a low-iron-content glass would seem to be the most adequate. With regard to the reflecting/concentrating materials, aluminium is the best option due to its low cost and high reflectivity in the solar UV spectrum on earth surface.

The original solar photoreactor designs for photochemical applications were based on line-focus parabolic-trough concentrators (PTCs) [5]. But there is a category of low concentration collectors, called Compound Parabolic Concentrators (CPCs), that are a good option for solar photochemical applications [9]. If the CPC is designed for an acceptance angle of +90 to −90°, all incident solar diffuse radiation can be collected (Fig. 4.2). The light reflected by the CPC is distributed all around the tubular receiver so that almost the entire circumference of the receiver tube is illuminated. They do so illuminating the complete perimeter of the receiver, rather than just the "front" of it, as in conventional flat plates. These concentrating devices have ideal optics, thus maintaining both the advantages of the PTC and static systems. The concentration factor ($R_C$) of a two dimensional CPC collector is given by Eq. 4.3. They can make highly efficient use of both direct and diffuse solar radiation, without the need for solar tracking. There is no evaporation of possible volatile compounds and water does not heat up. They have high optical efficiency, since they make use of almost all the available radiation, and high quantum efficiency, as they do not receive a concentrated flow of photons. Flow also can be easily maintained turbulent inside the tube reactor. Reports exist that provide excellent reviews of the needs towards the solar hardware for photocatalytic processes based on $TiO_2$ and photo-Fenton application including aspects of optics, geometry and reactor materials [8, 10–12].

$$R_{C,\text{CPC}} = \frac{1}{\sin \theta_a} = \frac{A}{2\pi r} \tag{4.3}$$

Nevertheless, the design procedure for a photocatalytic system requires the selection of a reactor, catalyst, reactor-field configuration (series or parallel), treatment-system mode (once-through or batch), flow rate, pH control, etc., so a plant has to be as versatile as possible and provide sufficient confidence. Usually, a photocatalysis plant is constructed with several solar collectors. All the modules are connected in series, but with valves that permit to bypass any number of them. All the connection tubes and valves are strongly resistant to chemicals, weatherproof and opaque, in order to avoid any photochemical effect outside of the collectors. The most important sensors required for the system are temperature, pressure and dissolved oxygen. A UV-radiation sensor must be placed in a position where the solar UV light reaching the photoreactor can be measured. Solar plants are frequently operated in a recirculating batch mode. In this situation, the fluid is continuously pumped between the reactor and a tank in which no reaction occurs, until the desired degradation is achieved. Each collector (see Fig. 4.2c) consists of Pyrex tubes (installed in the axis of the CPC) connected in series and mounted on a fixed platform tilted at local latitude (see Fig. 4.2d). The total volume ($V_T$) of the reactor is separated in: total irradiated volume (Pyrex tubes, $V_i$) and the dead reactor volume (tank + connecting tubes).

Solar UV is usually measured by a global UV radiometer mounted on a platform tilted at the same angle as the CPCs, which provides data in terms of incident W m$^{-2}$. This gives an idea of the energy reaching any surface in the same position with regard to the sun. With Eq. 4.4, evaluation of the data is possible, where $t_n$ is the experimental time, UV is the solar ultraviolet radiation measured during $\Delta t_n$, and $t_{30W}$ is a "normalized illumination time". In this case, time refers to a constant solar UV power of 30 W m$^{-2}$.

$$t_{30W,n} = t_{30W,n-1} + \Delta t_n \frac{\text{UV}}{30} \frac{V_i}{V_T}; \Delta t_n = t_n - t_{n-1} \tag{4.4}$$

## 4.4 Photocatalytic Degradation of Pesticides

In general, the types of pesticides that have been degraded by photocatalysis include a large number of structures [4]. Until now, the absence of total mineralisation has been observed only in $s$-triazine herbicides, for which the final product obtained was essentially 1,3,5-triazine-2,4,6, trihydroxy (cyanuric acid), which is, fortunately, nontoxic [13]. In photocatalysis, transformation of the parent organic compound is desirable in order to eliminate its toxicity and persistence, but the principal objective is to mineralise all pollutants. For chlorinated pesticides, $Cl^-$ ions are easily released in the solution and are the first of the ions appearing during the photocatalytic degradation. This could be interesting in a process, where photocatalysis would be associated with a biological treatment which is generally not efficient for chlorinated compounds. Nitrogen-containing pesticides are mineralised mostly into $NO_3^-$ and $NH_4^+$. Ammonium ions are relatively stable, and the proportion depends mainly on the oxidation stage of organic nitrogen and irradiation time [14]. The formation of $N_2$ in azo bounds can be accounted for by the same processes responsible for $NH_4^+$ formation [15]. $N_2$ evolution constitutes the ideal case for a decontamination reaction involving totally innocuous nitrogen-containing final product.

Organophosphorous pesticides produce phosphate ions. However, in the pH range used (usually > 4), phosphate ions remain adsorbed on $TiO_2$. This strong adsorption somewhat inhibits the reaction rate, though it is still acceptable. In photo-Fenton, phosphate sequestrates iron forming the corresponding non-soluble salt and retarding the reaction rate. Therefore, more iron is necessary when water containing phosphates is treated by photo-Fenton. Pesticides containing sulphur atoms are mineralised into sulphate ions. The release of $SO_4^{2-}$ can be accounted by an initial attack by a photo-induced $^{\bullet}OH$ radical. In all the studies the formation of $SO_4^{2-}$ was always observed and in most cases its stoichiometric formation was found in the final steps of the photoreaction when organic intermediates were still present. Initial rate was high indicating that $SO_4^{2-}$ ions are initial products, directly resulting from the initial attack on the sulfonyl group. Non-stoichiometric formation of sulphate ions is usually explained by a strong adsorption on the photocatalyst surface.

Sulphate, chloride and phosphate ions, especially at concentrations greater than 1 mM can reduce the rate due to the competitive adsorption at the photoactivated reaction sites of $TiO_2$. It has been also reported that photo-Fenton process efficiency is noticeably lowered in presence of chloride and sulfate ions [16]. There are two different reasons for this: (i) decreased generation of hydroxyl radicals because of the formation of chloro- and sulfato-Fe(III) complexes that affect the distribution and reactivity of the iron species, (ii) scavenging of hydroxyl radicals and formation of inorganic radicals ($Cl_2^{\bullet-}$ and $SO_4^{\bullet-}$), which are less reactive than $^{\bullet}OH$.

The effectiveness of degradation is not demonstrated only because the entire initial compound is decomposed. Reactants and products might be lost

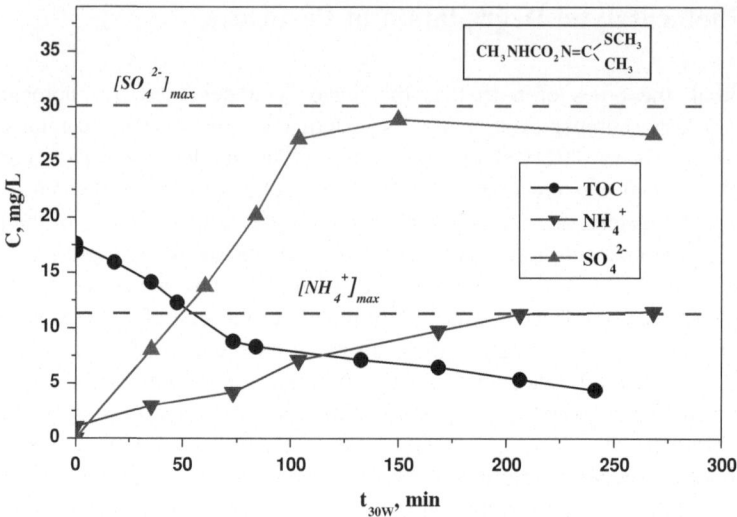

**Fig. 4.3** Degradation of methomyl (mineralisation and production of inorganic species). Methomyl structure is also shown

(evaporation, adsorption on reactor components, etc.) which introduces uncertainty in results. The mineralisation rate is determined by monitoring inorganic compounds, such as $CO_2$, $Cl_-$, $SO_4^{2-}$, $NO_3^-$, etc. When pesticides decompose, a stoichiometric increase in the concentration of inorganic anions is produced in the water treated (Fig. 4.3). For this reason, the analysis of these products of the reaction is of interest for the final mass balance.

Preliminary research is always required to assess pesticides treatments and optimise the best option for any specific problem, on a nearly case-by-case basis. In general, the types of pesticides which have been degraded include $s$-triazines, sulfonylureas, anilide and amide herbicides, carbamates, phenylureas, organophosphorous, organochlorines, chlorophenols, etc. Eq. 4.5 generally holds true for an organic compound of general formula $C_nH_mO_p$. In the case of pesticides containing halogens, Eq. 4.6 shows how the corresponding halide is formed. Sulphur is recovered as sulphate in sulphur containing pesticides according to Eq. 4.7.

$$C_nH_mO_p + \left(\frac{(m-2p)}{4} + n\right)O_2 \rightarrow nCO_2 + \frac{m}{2}H_2O \tag{4.5}$$

$$C_nH_mO_pX_q + \left(\frac{(m-2p)}{4} + n\right)O_2 \rightarrow nCO_2 + \frac{m-q}{2}H_2O + qHX \tag{4.6}$$

$$C_nH_mO_pS_r + xO_2 \rightarrow nCO_2 + yH_2O + zH_2SO_4 \tag{4.7}$$

The oxidation of carbon atoms into $CO_2$ is relatively easy. In general, at low reactant levels or for compounds which do not form important intermediates, complete mineralisation and reactant disappearance proceed with similar half lives, but at higher reactant levels where important intermediates occur, mineralisation is slower than the degradation of the parent compound. However, before photocatalytic treatment can be proposed as a general and trouble free method, it is required that the chemistry of various classes of pollutants under these conditions is known in detail. Since the chemistry of such processes is complex, careful analytical monitoring using different techniques is essential in order to control all transformation steps, to identify harmful intermediates and to understand and interpret the reaction mechanism. The assessment of pesticides disappearance in the early steps is not sufficient to ensure the absence of residual products because the photocatalytic treatment may give rise to a variety of organic intermediates which can themselves be toxic, and in some cases, more persistent than the original substrate [4]. From an analytical viewpoint, the task that entails the most difficulty is, without doubt, qualitative and quantitative evaluation of the intermediates or degradation products (DPs). As hydroxyl radicals are not selective in their attacks, numerous DPs form on the path towards complete mineralisation. There are five main types of DPs: (i) hydroxylated and dehalogenated products; (ii) products from the oxidation of the alkali chain, if it had one; (iii) products derived from the opening of the aromatic ring in aromatic contaminants; (iv) products of decarboxylation; (v) products of isomerization and cyclation. The chemical analysis of these complex reaction mixtures is difficult. However, a greater knowledge of the DPs originated would be necessary. It may be observed in Figure 4.4 that most of the DPs with high molecular weight appear after exposure to sunlight and reach their maximum concentration at short treatment time. From here on, they begin to decrease and carboxylic acids appear. Until now, the analyses of fragments resulting from the degradation of the aromatic ring have revealed formation of aliphatics (organic acids and other hydroxylated compounds), which explains why total mineralisation takes much longer than dearomatization, as mineralisation of aliphatics by photocatalysis is the slowest step [17].

To shorten phototreatment time is of major concern for the cost and energy efficiency benefits of the overall treatment process. Therefore, to investigate toxicity could be considered as a suitable overall indicator capable of giving information on the evolution of biocompatibility of the water solution contaminated with pesticides during the phototreatment in order to dispose to the environment or promote biotreatment. But due to the complexity of the studied process and the specificity and sensitivity of the toxicity test, this approach has to be considered and discussed with caution. Besides, a more detailed study of DPs as well as other inorganic species produced in addition with toxicity analyses should be achieved in order to improve the knowledge of the implicated degradation pathways and molecular interactions. Considering the removal of the initial contaminants on one hand and the mineralisation of the organic carbon on the other hand, two main categories of behaviours can be outlined. When the DPs demineralise shortly, toxicity usually decreases gradually in the course of the photodegradation. But when the reaction intermediates degradation takes a long time (after disappearance

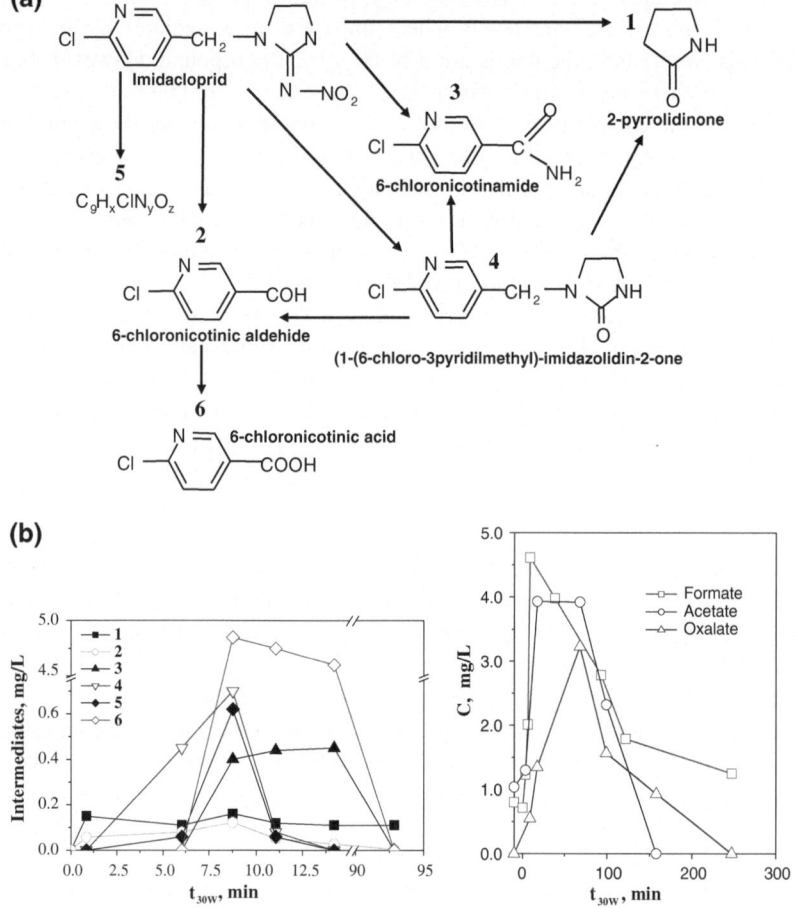

**Fig. 4.4 a** Scheme of the degradation pathway for Imidacloprid under solar photocatalytic treatment with photo-Fenton in water. **b** Formation and degradation of DPs

of the target contaminants), the level of toxicity is not predictable. However at the end, the toxicity tends to decrease.

Toxicity assessment of a chemical using a single species test reflects the sensitivity of that test only; it may overestimate or underestimate the potential toxicity for that particular substance. Accordingly, recent research has focused on the development of representative, cost-effective and quantitative test bioassays, which can detect different effects using a variety of endpoints [18]. Figure 4.5 shows the evolution of representative toxicity curves (% inhibition) for bioassays performed during solar photocatalytic experiments. Chemicals added to the water for photocatalysis were removed prior to bioassay and the pH was neutralised. TiO$_2$ was removed by filtration, H$_2$O$_2$ by quenching with catalase and iron by

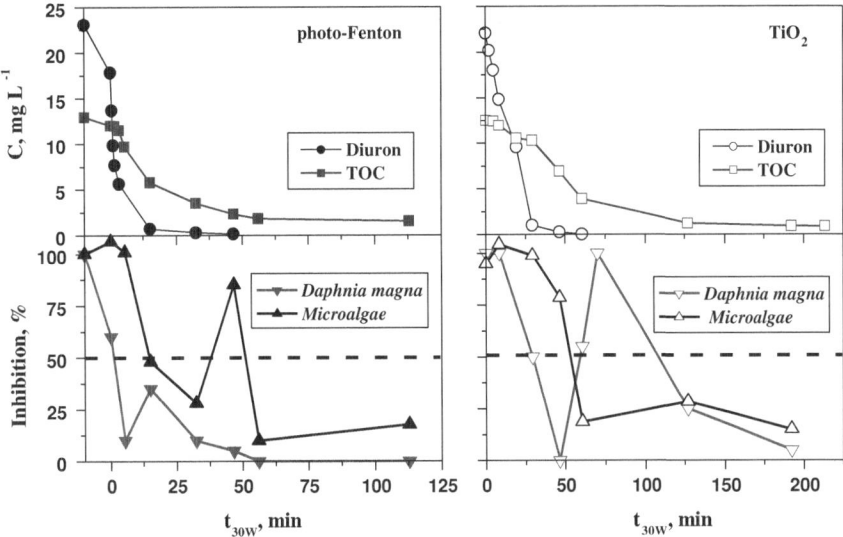

**Fig. 4.5** Degradation of diuron and TOC and evolution of toxicity

coagulation and filtration after neutralising the samples. These are not the results of a single toxic response, because in the experiments they are affected not only by the parent compound, but also by the presence of other intermediate compounds produced during its photodegradation. *Daphnia magna* was biochemically the most complex test system and also the most sensitive. *Selenastrum capricornotum* (microalgae) behaviour is different. At the beginning, diuron was toxic (100% inhibition) for both microorganisms. A very toxic intermediate (at least for *Selenastrum capricornotum*) is formed after the complete disappearance of diuron (at around 35 min by photo-Fenton and 75 min by $TiO_2$). This intermediate (or intermediates) may also be removed, because toxicity is reduced after a few minutes more of photo treatment. It can therefore be deduced that a very toxic (highly toxic at very low concentrations) unknown DP is formed at the end of photo treatment, but is degraded after a few minutes more of photo treatment. This is a clear demonstration that complete control of the degradation process must be achieved in order to guarantee overall AOP treatment. Toxicity could also be an alternative indicator for biodegradation assessment of partially photo-treated wastewaters [19]. Overall, acute toxicity testing has been shown to represent dynamics and efficiency of photo treatment. Very often toxicity changes continuously during the treatment, and therefore, toxicity evaluation is not a suitable way to determine the moment when biodegradability is most enhanced. However, reduced toxicity results are indicative of an extended biodegradability achieved during the process. These assays must therefore be complemented with biodegradability studies. Thus even if it cannot provide a reliable biodegradability assessment by itself, toxicity can help identifying samples to be tested by biodegradability assessment methods, which are quite time-consuming.

## 4.5 Photocatalytic Degradation of Wastewater Containing Pesticides

Several ways exist to enhance the performance of pesticides treatment by an AOP. The first possibility is to position the AOP in a sequence of physical, chemical and biological treatments (not necessarily in this order). Many times such a treatment approach will at least involve an AOP step and a biological treatment step. Either way, putting AOP or biological treatment first in the treatment train, the global objective of minimal costs will closely resemble minimising the treatment degree in the AOP and maximising the treatment in the biological treatment, because of the large differences in costs of the two different treatments [20]. The key issue is the correct design of the process, so that the process will be best in terms of overall economic and ecological performance.

A second option would be the real integration with another process, which may again be of physical, chemical or biological nature and we will first review several proposed possibilities for process integration.

Several authors have proposed the direct interaction of the oxidative mechanisms of photocatalysis with other chemical processes [21]. A series of integration approaches exist for the simultaneous application of photocatalysis and physical separation processes (e.g. activated carbon, nanofiltration [22] or membrane distillation [23]. Furthermore, there exist plenty of examples, which focus on the sequential combination of photo-Fenton treatment and biological treatment (aerobic in most cases). A general approach for the development of the combined treatment for wastewater containing pesticides will be discussed here.

The use of AOPs as a pre-treatment step to enhance the biodegradability of waste water containing pesticides can be justified if the resulting intermediates are readily degradable by microorganisms in further biological treatment. Today combined photo-assisted AOP and biological processes are gaining in importance as treatment systems [20, 24–32] as one of the main urban waste water treatment obligations imposed by European Union Council Directive 91/271/EEC is that waste water collecting and treatment systems (generally involving biological treatment), must be in place in all agglomerations since 31st December 2005 [33]. This means that nowadays, provided that the regulations have been implemented, AOP plants developed in the EU can discharge pre-treated waste water into a nearby conventional biological treatment plant. The same is true for many other locations all over the world.

When preliminary chemical oxidation is applied in a combination treatment line, sometimes its effect is insignificant or even harmful to the properties of the original effluent, even though it is conceptually advantageous. There are several reasons for this, the most common of which are: (i) formation of stable intermediates which are less biodegradable than the original molecules; (ii) lack of selectivity for preferential attack on the more bioresistant fractions of the wastewater during chemical treatment; (iii) poor selection of treatment conditions as, for example, excessive pre-oxidation can lead to generation of an effluent with too little metabolic value for the microorganisms; (iv) too much oxidant and/or catalyst used for oxidation remained in the pre-treated wastewater. Compounds such as ozone and hydrogen peroxide (both known as biocides), metals, metal oxides and metal salts (catalysts in many processes), are normally toxic to microorganisms.

These limitations underline the need to establish a step-by-step research methodology which takes these effects into account, because operating conditions effect on the original properties of the pre-treatment stream (contact time, oxidant and/or catalyst type, dose and toxicity, temperature, etc.) must be known. Such studies must employ analytical tools to infer the reaction mechanisms, pathway and kinetics, evaluate the effect of the chemical pre-treatment on toxicity and biodegradability, the effect of cations and anions in the wastewater matrix, and the application of various techniques for determining biodegradability and toxicity.

TOC (or COD) as a general parameter of wastewater treatment should always be known. If the wastewater is not biodegradable and TOC is high (>100 mg/L) AOP pre-treatment before biotreatment should be envisaged (AOP/BIO). After the treatment the effluent quality has to be checked, to decide if it complies with legal requirements for effluent discharge. If the wastewater is not biodegradable but TOC is low (<100 mg/L), one should design the appropriate AOP treatment but without a subsequent biotreatment, because such a low TOC would not produce pre-treated effluent (this means, with lower TOC) suitable for a biotreatment. Very often this wastewater could be disposed to the environment after the AOP treatment or, which is more convenient, to a public sewage treatment system for polishing it.

To develop and optimise coupling strategy (AOP/BIO) is truly multidisciplinary and requires knowledge of the biological and the chemical process. A series of analytical parameters needs to be measured ranging from chemical sum parameters (total organic carbon or chemical oxygen demand), chromatographic methods (HPLC–UV to quantify specific contaminants of interest), acute toxicity tests (typically various, e.g. *Vibrio Fischeri* and *Daphnia Magnae*) to biodegradability tests ($BOD_5$, Zahn-Wellens test, respirometry). This whole series of analytical parameters will satisfy the needs for engineering purposes to design the coupling strategy. In coupled systems, the AOP pretreatment is meant to modify the structure of pesticides by transforming them into less toxic and easily biodegradable intermediates, which allows the subsequent biological degradation to be achieved in a shorter time and in a less expensive way.

These requirements, together with information concerning the evolution of toxicity and biodegradability of the phototreated solutions, allow the determination of an optimal phototreatment time, which corresponds to the best cost-efficiency compromise. However, if the fixed pretreatment time is too short, the intermediates remaining in solution could still be structurally similar to initial biorecalcitrant compounds and therefore, non-biodegradable.

In the OECD guidelines biodegradation tests are divided into three principal categories: tests for ready biodegradability, tests for inherent biodegradability and simulation tests. Tests for inherent biodegradability such as the Zahn-Wellens

(Z-W) procedure is the most appropriate method for biodegradation assessment of partially photodegraded solutions of pesticides. But this analytical tool is quite time-consuming, typically between a few days in the case of quick evidence of biode-gradability (i.e. samples with ready biodegradability) and in the case of a continuing negative test response the test must be prolonged for 4 weeks, which is the test duration according to the standard protocol. Therefore, to limit the amounts of samples to be processed by the Zahn-Wellens test (Z-W), we propose as an indicator of partially phototreated waters the complementary use of acute toxicity techniques, which yield a comparably quick response. A series of acute toxicity tests are avail-able [34], but an interesting cross between ready biodegradability tests and acute toxicity tests are short-term respirometric tests performed with activated sludge [35]. Figure 4.6 shows toxicity results on samples taken at different stages of the photo-Fenton as a percentage of bacteria inhibition when exposed to samples for 30 min. Inhibition decreased from 80% for the non-treated pesticide wastewater to 50% after phototreatment (final TOC of 50 mg/L), but in between there were stages when toxicity was lower. Interestingly, toxicity reduction was more pronounced in those samples where complete elimination of the active ingredients was achieved. With longer phototreatment treatment, toxicity increased slightly, presumably due to the formation of toxic end products. The most important result of the toxicity tests was that it changed continuously during the treatment indicating that biodegradability should change dramatically during the process. These assays must therefore be complemented with biodegradability studies, as stated in the following paragraph.

The Z-W test was performed on six samples taken at different stages of the photo-Fenton process to check their aerobic biodegradability. Sample 1 is the original pesticide wastewater, while 2 has been subjected to mild phototreatment and both of them contain active ingredients. Other samples are free of pesticides. Therefore, as expected, pesticides were nonbiodegradable. As shown in Fig. 4.6, samples 1 and 2 were hardly biodegradable, with only 50 and 60% biodegradability after 28 days of biotreatment, respectively. On the other hand, all samples without pesticides reached at least 70% of biodegradability in 9 days (TOC < 175 mg/L, see Fig. 4.6). The continuous enhancement of biodegradability fits well with reduction in toxicity. Feasible combination of phototreatment and a biological treatment was demon-strated but in order to optimise the combined system and reduce costs, the photo-Fenton process should be as short as possible, so as much TOC as possible must be eliminated by the biotreatment, which has been demonstrated to be more cost efficient and environmentally friendly [36].

## 4.6 Concluding Remarks

The proposed technology could also be applicable to other organic hazardous contaminants, such as solvents, detergents and a variety of industrial chemicals, which are capable of deep penetration into the soil and reach groundwater. Solar AOPs have the advantage over other AOPs of using sunlight and having as its main characteristic that it is an environmentally friendly technology.

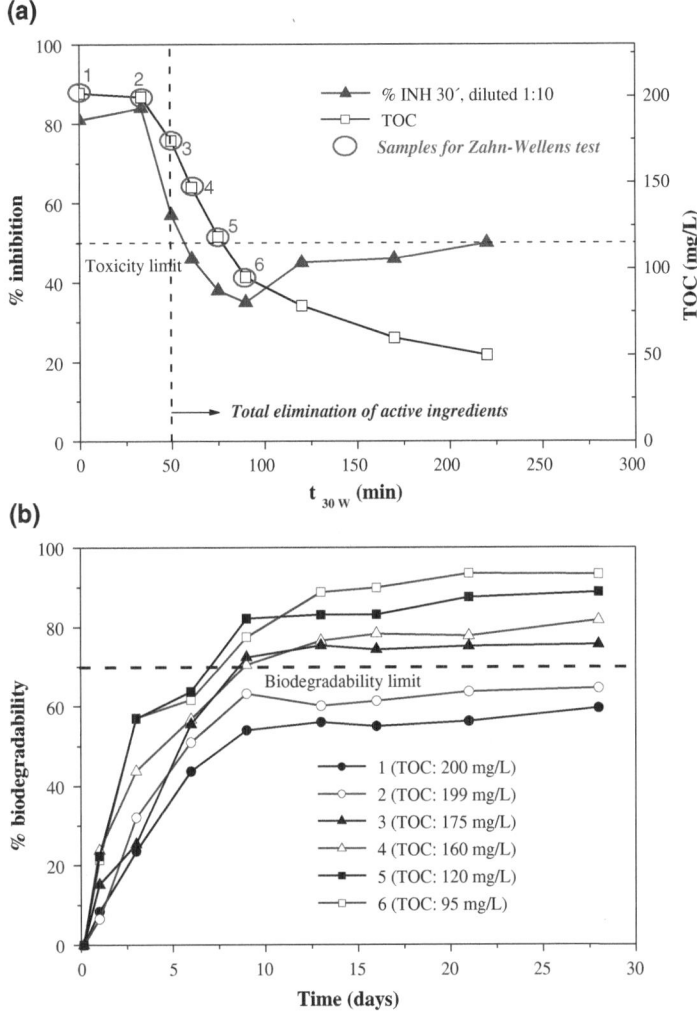

**Fig. 4.6** Treatment of wastewater containing five commercial pesticides commonly used in intensive agriculture Vydate (10% Oxamyl), Metomur (20% Methomyl), Couraze (20% Imidacloprid), Ditimur-40 (40% Dimethoate) and Scala (40% Pyrimethanil) by solar photo-Fenton. **a**. Percentage of *Vibrio fischeri* inhibition after 30 min exposure to samples partially treated and **b**. Z-W biodegradability analysis of selected (1–6) samples

Contaminant treatment, in its strictest meaning, is the complete mineralisation (TOC = 0) of the contaminants, but when feasible, biological treatment is the cheapest treatment and also the most compatible with the environment. Therefore, biologically recalcitrant compounds could be treated with photocatalytic technologies until biodegradability is achieved, later transferring the water to a conventional biological plant. Such a combination reduces treatment time and optimises the overall economics, since the solar detoxification system can be significantly smaller. Due to the kinetic mechanism, the first part of the photo-catalytic process is the quickest. As it can be seen in the figures presented, when the active component disappears, TOC remains for a long time. Therefore, the use of AOPs as a pre-treatment step can be justified if the intermediates resulting from the reaction (more oxidised compounds as carboxylic acids, alcohols, etc.) are readily degraded by microorganisms. The feasibility of such a photocatalytic-biological process combination must always be assessed, because it could provide an important cost reduction by reducing the size of the necessary solar collector field. It must be taken into account that, as with most solar systems, economics of the water detoxification systems are dominated by their capital cost.

Determining the toxicity of the water, at different stages of AOP treatment, using different microorganisms is another way to decrease AOP operating costs. In this case, biocompatibility with the environment can be stated. Toxicity testing of the photocatalytically treated wastewater is therefore necessary, particularly when incomplete degradation is planned. Recently, the use of acute toxicity bioassays has meant an important improvement in the evaluation of AOPs because of their reproducibility, adequate format for quick analysis, short analysis time, as well as well-defined analytical protocols.

**Acknowledgments** The authors wish to thank the Spanish Ministry of Science and Innovation for financial support under EDARSOL project (reference: CTQ2009-13459-C05-01).

# References

1. Andreozzi R, Caprio V, Insola A, Martota R (1999) Advanced oxidation proceses (AOP) for water purification and recovery. Catal Today 53:51–59
2. Bahnemann D (2004) Photocatalytic water treatment: solar energy applications. Sol Energy 77:445–459
3. Greenhalgh R (1980) Definition of persistence in pesticide chemistry. Pure Appl Chem 52:2563–2566
4. Konstantinou IK, Albanis TA (2003) Photocatalytic transformation of pesticides in aqueous titanium dioxide suspensions using articficial and solar light: intermediates and degradation pathways. Appl Catal B Environ 42:319–335
5. Dillert R, Cassano AE, Goslich R, Bahnemann D (1999) Large scale studies in solar catalytic wastewater treatment. Catalysis Today 54:267–282
6. Herrmann JM (2010) Photocatalysis fundamentals revisited to avoid several misconceptions. Appl Cat B Environ 99:461–468
7. Pignatello JJ, Oliveros E, MacKay A (2006) Advanced oxidation processes for organic contaminant destruction based on the Fenton reaction and related chemistry. Critical Rev Environ Sci Technol 36:1–84
8. Malato S, Fernández-Ibáñez P, Maldonado MI, Blanco J, Gernjak W (2009) Decontamination and disinfection of water by solar photocatalysis: recent overview and trends. Catal Today 147:1–59
9. Ajona JA, Vidal A (2000) The use of CPC collectors for detoxification of contaminated water: design, construction and preliminary results. Sol Energy 68:109–120

10. Malato S, Blanco J, Vidal A, Richter C (2002) Photocatalysis with solar energy at a pilot-plant scale: an overview. Appl Catal B Environ 122:137–149
11. Malato S, Blanco J, Maldonado MI, Fernández P, Alarcon D, Collares M, Farinha J, Correia J (2004) Engineering of solar photocatalytic collectors. Sol Energy 77:513–524
12. Malato S, Blanco J, Alarcón DC, Maldonado MI, Fernández P, Gernjak W (2007) Photocatalytic decontamination and disinfection of water with solar collectors. Catal Today 122:137–149
13. Pelizzetti E, Maurino V, Carlin V, Pramauro E, Zerbinati O, Todato ML (1990) Photocatalytic degradation of atrazine and other s-triazine herbicides. Environ Sci Technol 24:1559–1565
14. Calza P, Pelizzetti E, Minero C (2005) The fate of organic nitrogen in photocatalysis: an overview. J Appl Electrochem 35:665–673
15. Konstantinou IK, Albanis TA (2004) TiO2-assisted catalytic degradation of azodyes in aqueous solutions: kinetic and mechanistic investigations. Appl Catal B Environ 49:1–14
16. Le Truong G, De Laat J, Legube B (2004) Effects of chloride and sulfate on the rate of oxidation of ferrous ion by $H_2O_2$. Water Res 38:2384–2394
17. Burrows HD, Cable L, Santaballa M, Steenken JA (2002) Reaction pathways and mechanisms of photodegradation of pesticides. J Photochem Photobiol B Biol 67:71–108
18. Davoren M, Fogarty AM (2004) A test battery for the ecotoxicological evaluation of the agri-chemical Environ. Environ Ecotoxicol Environ Saf 59:116–122
19. Lapertot M, Ebrahimi S, Oller I et al (2008) Evaluating Microtox as a tool for biodegradability assessment of partially treated solutions of pesticides using $Fe^{3+}$ and $TiO_2$ solar photo-assisted processes. Ecotoxicol Environ Saf 69:546–555
20. Mantzavinos D, Psillakis E (2004) Enhancement of biodegradability of industrial wastewaters by chemical oxidation pre-treatment. J Chem Technol Biotechnol 79:431–454
21. Gogate PR, Pandit AB (2004) A review of imperative technologies for wastewater treatment II: Hybrid methods. Adv Environ Res 8:553–597
22. Augugliano V, Litter M, Palmisano L, Soria J (2007) The combination of heterogeneous photocatalysis with chemical and physical operations: a tool for improving the photo process performance. J Photochem Photobiol C Photochem Rev 7:123–144
23. Mozia S, Tomaszewska M, Morawski AW (2007) Photocatalytic membrane reactor (PMR) coupling photocatalysis and membrane distillation—effectiveness of removal of three azo dyes from water. Catal Today 129:3–8
24. Pulgarín C, Invernizzi M, Parra S, Sarria V, Polaina R, Péringer P (1999) Strategy for the coupling of photochemical and biological flow reactions useful in mineralization of biocalcitrant industrial pollutants. Catalysis Today 54:341–352
25. Wang XJ, Chen SL, Gu XY, Wang KY, Qian YZ (2008) Biological aerated filter treated textile washing wastewater for reuse after ozonation pre-treatment. Wat Sci Technol 58:919–923
26. Badawy MI, Wahaab RA, El-Kalliny AS (2009) Fenton-biological treatment processes for the removal of some pharmaceuticals from industrial wastewater. J Hazar Mat 167:567–574
27. Basha CA, Chithra E, Sripriyalakshmi NK (2009) Electro-degradation and biological oxidation of non-biodegradable organic contaminants. Chem Eng 149:25–34
28. Di Iaconi C, Ramadori R, Lopez A (2009) The effect of ozone on tannery wastewater biological treatment at demonstrative scale. Bioresour Technol 100:6121–6124
29. Lafi WK, Shannak B, Al-Shannaga M, Al-Anbera Z, Al-Hasa M (2009) Treatment of olive mill wastewater by combined advanced oxidation and biodegradation. Sep Pur Technol 70:141–146
30. Wang XJ, Chen S, Gu X, Wang K (2009) Pilot study on the advanced treatment of landfill leachate using a combined coagulation. Fenton oxidation and biological aerated filter process waste management 29:1354–1358
31. Mascolo G, Laera G, Pollice A, Cassano D, Pinto A, Salerno C, Lopez A (2010) Effective organics degradation from pharmaceutical wastewater by an integrated process including membrane bioreactor and ozonation. Chemosphere 78:1100–1109

32. Mendoza-Marín C, Osorio P, Benítez N (2010) Decontamination of industrial wastewater from sugarcane crops by combining solar photo-Fenton and biological treatments. J Hazar Mat 177:851–855
33. Commission European (2004) Implementation of council directive 91/27/EEC of 21 May 1991 concerning urban waste water treatment, as amended by commission directive 98/15/EC of 27 Feb 1998. Commission of the European Communities, Brussels
34. Rizzo L (2011) Bioassays as a tool for evaluating advanced oxidation processes in water and wastewater treatment. Wat Res 45:4311–4340
35. Amat AM, Arques A, García-Ripoll A et al (2009) A reliable monitoring of the biocompatibility of an effluent along an oxidative pre-treatment by sequential bioassays and chemical analyses. Wat Res 43:784–792
36. Muñoz I, Peral J, Ayllón JA, Malato S et al (2007) Life cycle assessment of a coupled advanced oxidation-biological process for wastewater treatment comparison with granular activated carbon adsorption. Environ Eng Sci 24:638–651

# Chapter 5
# Removal of Pharmaceutics by Solar-Driven Processes

Antonio Arques and Ana Maria Amat

**Abstract** Pharmaceuticals in wastewater constitute an increasing environmental concern as a consequence of human consumption, veterinary use and industrial production of these compounds. Although they are commonly found at low concentration, their effect on human health and environment is not yet established. Oxidative photochemical methods using sunlight constitute promising alternatives to non-efficient conventional treatments. Titanium dioxide and photo-Fenton have been employed to remove a number of pharmaceuticals from water. Although most experiments involve model compounds at relatively high concentrations, some information is available on the treatment of real effluents or to determine the effect of the water-matrix on the processes. In view of their practical application, the processes have been scaled-up to pilot plant and preliminary economic evaluations are available. Finally, photolysis of pharmaceuticals, although cannot be considered as an actual treatment technology, is of paramount importance because of its contribution to the self-cleaning of aqueous ecosystems.

**Keywords** Pharmaceuticals · Wastewater · Solar processes · Photocatalysis · Photo-Fenton · Titanium dioxide · Detoxification · By-products · Pilot plant · Photolysis · Self-cleaning

A. Arques (✉) · A. M. Amat
Departamento de Ingeniería Textil y Papelera,
Universidad Politécnica de Valencia,
Campus de Alcoy Plaza Ferrándiz y Carbonell s/n,
03801, Alcoy, Spain
e-mail: aarques@txp.upv.es

A. M. Amat
e-mail: aamat@txp.upv.es

G. Lofrano (ed.), *Emerging Compounds Removal from Wastewater*,
SpringerBriefs in Green Chemistry for Sustainability,
DOI: 10.1007/978-94-007-3916-1_5, © Arques, Amat 2012

## 5.1 Introduction

In recent years, the presence of an increasing number of chemical substances at low concentrations in surface waters and wastewaters because of human activities has become a serious environmental concern as they might represent a threat for natural ecosystems and a limitation for the potential re-use of wastewaters. Perfluorinated compounds, pharmaceuticals, hormones, endocrine disruptors, drinking water and swimming pool disinfection byproducts, sunscreens, flame retardants, algal toxins, dioxane, pesticides or nanomaterials are examples of these chemicals and they are commonly classified as emerging pollutants (EPs) [1, 2].

Among the emerging pollutants, pharmaceuticals constitute a major problem as increasing amounts of these drugs are released into the environment. Main sources of these chemicals in water are excretion of non-metabolized drugs by humans or animals, and flushing of unused medication or discharge of wastewaters from pharmaceutical industry [3]. As a consequence of this, analgesics, antibiotics, anticonvulsants, cytostatics, hormones, $\beta\beta$-blockers, antihypertensives, antihistamines, lipid regulators, stimulants or fragrances have been detected in urban wastewaters entering wastewater treatment plants at concentrations ranging from a few ng/l until more than 100 µg/l [4].

Conventional methods for wastewater treatment are not always suitable to remove pharmaceuticals and hence, they are systematically found at the effluents of wastewater treatment plants and then released into the environment or re-used for other human activities [5]. These chemicals show resistance to microbial biodegradation, and chronic exposure might produce adverse effects on aquatic life; some drugs or their metabolites have been even found to reach significant concentrations in surface and drinking water which might constitute a risk for human health [3, 6]. For this reason alternative methods are needed to deal with this concern. Among the treatments that have been tested are coagulation-flocculation, bioprocesses based on membrane biological reactors or constructed wetlands, nanofiltration, reverse osmosis, ozonation or chlorination [3, 4, 7].

## 5.2 Solar-Based Advanced Oxidation Methods

The use of photochemical methods involving sunlight for wastewater treatment has deserved increasing attention from researchers in recent years; these treatments are able to generate highly oxidizing species upon solar irradiation in the presence of a photocatalyst [8]. Titanium dioxide is the most widely employed photocatalyst [9]; it is a solid semiconductor that upon UV irradiation generates highly energetic electrons in the conduction band and holes in the valence band (Fig. 5.1). The holes behave as electron acceptors, which can oxidize the substrate via an electron transfer process between the adsorbed pollutant and $TiO_2$ (Eq. 5.1); in addition, other reactive species, such as hydroxyl radicals can be formed (Eq. 5.2); on the

**Fig. 5.1** Scheme of the band diagrams for TiO$_2$: upon irradiation, an electron is promoted from the valence band to the conduction band, generating a hole (h$^+$) in the valence band. Some reactions of holes and electrons are also given (Eqs. 5.1–5.3)

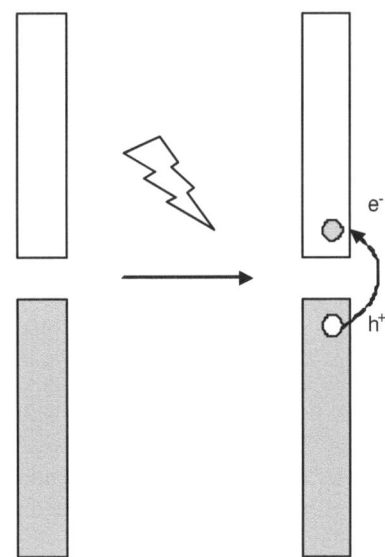

other hand, recombination of electron and holes results in a loss of efficiency of the process (Eq. 5.3).

$$h^+ + \text{pollutant} \rightarrow \text{oxid. pollutant} \tag{5.1}$$

$$e^- + H_2O_2 \rightarrow OH^- + \cdot OH \tag{5.2}$$

$$e^- + h^+ \rightarrow \text{recombination} \tag{5.3}$$

The photocatalyst is added in slurry, although some experiments can be found using supported TiO$_2$ (see Sect. 5.4). Different amounts of TiO$_2$ have been tested to treat water polluted with pharmaceuticals, generally in the range 0.2–2.0 g/l; higher amounts of photocatalyst are inefficient because photons are not able to reach the inner particles, which are shadowed by the outer ones. In some cases, hydrogen peroxide has been added in order to enhance the effect of the photocatalyst: this compound is able to react with excited electrons found in the conduction band of the semiconductor (see Eq. 5.2); this prevents recombination of the excited electrons with the holes of TiO$_2$, and improves the efficiency of the photocatalyst; furthermore, highly reactive·OH are formed in the process.

Other solid semiconductors have been tested for the treatment of pharmaceuticals. For instance ZnO has been reported to catalyze the photo-oxidation of some compounds such as carbamazepine [10] or the antibiotic tetracycline [11]. In fact, faster removal of tetracycline was measured with ZnO under optimized conditions (basic pH), although the major drawback of this material is that suffers corrosion at lower pH values.

A promising alternative to semiconductor-based solar photocatalysis is the photo-Fenton process; Fenton process consists in a mixture of iron salts and hydrogen peroxide. In this process, iron is able to catalyze decomposition of hydrogen peroxide into highly reactive hydroxyl radicals; the most important steps of the process are given by Eqs. 5.4 and 5.5 [12]. Although the Fenton reaction occurs in the dark, it is greatly enhanced by irradiation. Photo-Fenton is not considered a photocatalytic method by all authors, as stoichiometric amounts of peroxide are needed; however, there is no doubt that it is a solar-driven oxidative process and hence, it must be included in this chapter.

$$Fe^{2+} + H_2O_2 \rightarrow Fe^{3+} + OH^- + OH \qquad (5.4)$$

$$Fe(OH)^{2+} + h\nu \rightarrow Fe^{2+}OH \qquad (5.5)$$

For the remediation of water-containing pharmaceuticals the iron concentrations used, generally added as ferrous sulfate, are commonly in the range of 10–50 mg/l. As hydrogen peroxide is consumed in the reaction, the added amount of this reagent is strongly dependent on the amount of organic matter and on the intensity of treatment that is required.

Maybe one of the major drawbacks of the photo-Fenton process is that a highly acidic medium is required, as the optimal pH is 2.8. Hence, the effluent has to be acidified to reach this value and then neutralization is required before discharge. However, important efforts are being devoted to develop photo-Fenton processes under milder conditions.

## 5.3 Treatment of Model Compounds

The treatment of pharmaceuticals with titanium dioxide or photo-Fenton has been assayed using model compounds. For this purpose the concentrations of chemicals employed (typically mg/l) are several orders of magnitude above the concentrations found in natural waters or at the outlet of wastewater treatment plants (µg/l or ng/l), as higher concentrations can be submitted to a more accurate analyses (see Fig. 5.2 for the range of concentrations of pharmaceuticals found in different media). Irradiations can be performed using lamps emitting in the UVA range of the spectrum, with solar simulators or under solar irradiation.

The concentration of the pollutants has been usually determined by HPLC analysis and kinetic data can be fitted to a pseudo-first-order law or to Langmuir–Hinshelwood kinetic model in the case of TiO$_2$. Among the pharmaceuticals whose elimination with real or simulated sunlight has been studied are analgesics, stimulants, anti-inflammatories, anticonvulsants, antibiotics, steroid hormones, $ID = "IEqb" > \beta$-blockers, and cholesterol-lowering statins. A non-exhaustive list of these chemicals is given in Table 5.1.

Primary removal of the pollutants can be achieved after relative short irradiation periods, ranging from a few minutes to some hours, depending on the initial concentration of sample, the irradiation source, the sample matrix and the

**Fig. 5.2** Typical range
of concentrations of
pharmaceuticals in different
media: synthetic water
prepared for laboratory
experimentation,
pharmaceutical wastewaters,
influent and effluent of
wastewater treatment plants
and natural ecosystems

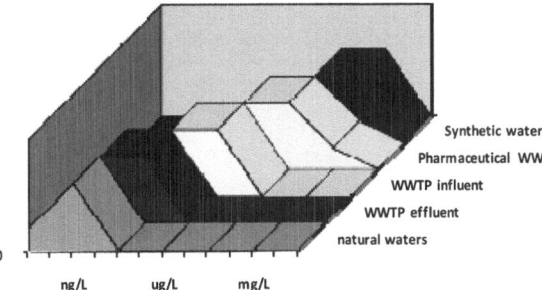

treatment that has been used; for example, photo-Fenton has been demonstrated to be faster than $TiO_2$. However, total abatement of the drug does not guarantee decontamination of the effluent, as other organic by-products are formed in the oxidative process. Mineralization of organic matter can be evaluated by means of Dissolved Organic Carbon (DOC) analysis. Although significant decrease in DOC has been observed in most cases, this process is slower than pharmaceuticals removal. Other gross-parameter that can be employed to monitor the process is Chemical Oxygen Demand (COD).

By-products formed in the photochemical process have been identified using gas or liquid chromatography equipped with a mass detector (GC–MS or LC–MS). These analyses have been performed with some pharmaceuticals. For example, intermediates formed in the $TiO_2$-mediated photo-oxidation of chemicals such as carbamazepine [10], ciprofloxacine [13], fluoroquinolone [14], atenolol [15], trimethoprim [16] or lovastatin, pravastatin and sinvastatin [17]. In the case of photo-Fenton, by-products generated from ampicillin [18] have also been studied. In some cases a reaction mechanism has been proposed. For instance, Trovó et al. proposed a mechanism for the photo-oxidation of sulfamethoxazole by means of a solar photo-Fenton process (see Fig. 5.3); the identification of by-products was based on a sophisticated Liquid Chromatography Electrospray Time of Flight Mass Spectrometry (LC-ESI-TOF–MS) analysis [19].

Detoxification assessment of the effluents is a necessary step prior discharge in fresh waters or application of a consecutive biological treatment if mineralization has not been accomplished. Toxicity of solutions containing the pharmaceuticals has been monitored along the solar-driven process according to different bioassays, such as inhibition of the luminescence of *Vibrio fischeri* bacteria [16, 17, 20], on *Vibrio quing-haiensi* [21], or the mobility of *Daphnia magna* [15, 22]. Elimination of the parent pollutant did not always result in a decrease in the toxicity; in some cases, an enhanced toxicity has been observed at the early stages of the reaction, which has been attributed to the formation of highly toxic intermediates. For instance, the elimination of diclofenac photocatalyzed by $TiO_2$ resulted in an enhanced toxicity according to the inhibition of the mobility of *D. magna* [22]. However, longer periods of irradiation have been usually demonstrated to be able to destroy these compounds and to form more biocompatible products; hence,

**Table 5.1** Classification and chemical structures of some pharmaceuticals whose elimination by solar-driven photochemical processes has been studied

| Analgesic | | | |
|---|---|---|---|
| | Acetaminophen | Antipyrine | Paracetamol |

| Stimulant | |
|---|---|
| | Caffeine |

| Anti-inflamatory | | | |
|---|---|---|---|
| | Diclofenac | Ibuprofen | Ketorolac |

| Anticonvulsant | |
|---|---|
| | Carbamazepine |

| Antibiotics | | | |
|---|---|---|---|
| | Flumequine | Ofloxacin | Triclosan |
| | Amoxicillin | Sulfamethoxazole | Erythromycin |

| Steroid hormone | |
|---|---|
| | Progesterone |

(continued)

**Table 5.1** (continued)

| β- blocker | Atenolol | Propranolol |
|---|---|---|
| | Metropolol | |
| Cholesterol-lowering statin | | |
| | Lovastatin | Pravastatin | Simvastatin |

detoxification occurs. It is interesting to point that in the case of antibiotics, remaining antibiotic activity has been monitored by adding the photo-treated mixture to agar plates which were inoculated with micro-organisms such as *Staphylococcus aureus* or *E. coli* [11, 23] and the effect of the mixture on the microorganisms is evaluated according to the size of the inhibition halo they produce.

Biodegradability has also been determined in some cases; however, this parameter is only interesting when solar photocatalysis is designed as a pretreatment of a biological process. This is not the case of the effluents of wastewater treatment plants, where photocatalysis is a tertiary treatment but these tests could be of interest for effluents of pharmaceutical industries. For instance, an enhancement of the biodegradability was determined during the treatment of paracetamol at relatively high concentration (157 mg/l) by means of a photo-Fenton process using a bioassay with the bacterium *Pseudomonas putida*, which was based on DOC consumption [24]; the $BOD_5$/COD ratio has been employed in the photo-Fenton treatment of metoprolol [20] and a long-term biodegradability assays, namely Zahn-Wellens test, has been used to evaluate the biocompatibility of nalidixic acid after solar photo-Fenton [25].

An important number of operational variables have a remarkable influence on the performance of the solar treatment. In order to optimize the process, statistical methods, generally based in a surface response methodology, have been applied. For instance, the effect of pH and $TiO_2$ concentration has been determined in the degradation of flumequine using this methodology [26]. In the case of photo-Fenton, variables such as pH and iron concentration of $H_2O_2$ have been optimized for the elimination of ampicillin [27].

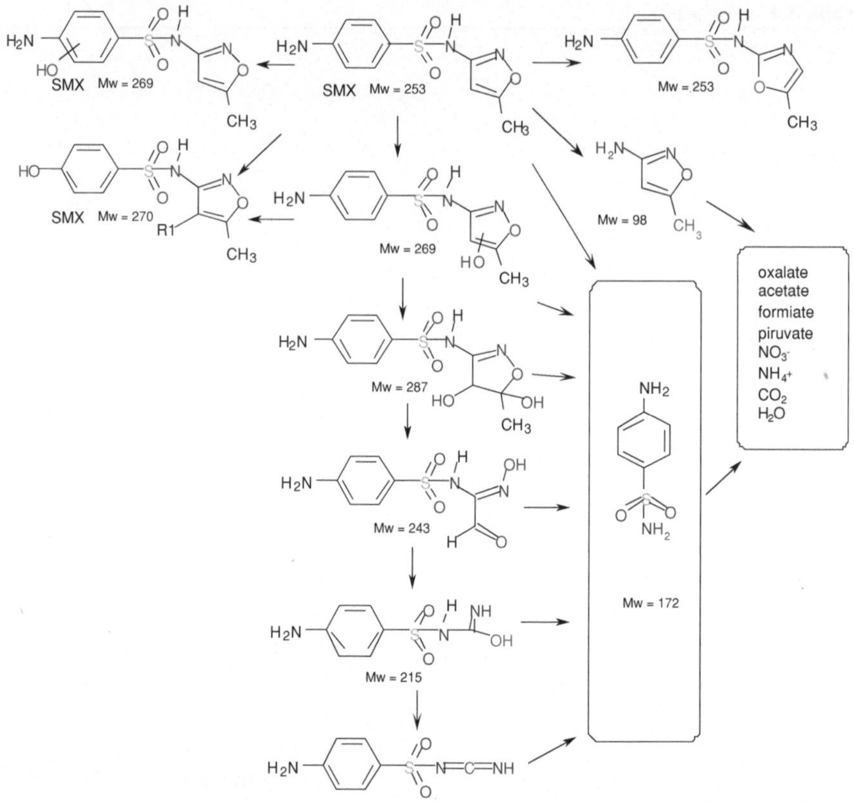

**Fig. 5.3** Reaction pathway for the photochemical oxidation of sulfamethoxazole in the presence of $TiO_2$ under sunlight irradiation

## 5.4 Toward Real Applications. Green Aspects of the Technology

As stated above, solar-driven processes can be considered as a green technology as they employ sunlight as irradiation source, avoiding the use of highly energy consuming UV-lamps; furthermore hazardous chemicals are not involved in the process as the oxidative species are generated "in situ"; finally robust set-ups are employed, which results in low maintenance costs and potential installation in isolated rural areas. However, results of laboratory experiments cannot be extrapolated straightforward to real scenarios; hence, extra research is still required to solve important practical problems.

In most of the experiments involving model compounds, distilled water has been used as solvent. However, the effect of the matrix on the photochemical reaction has been investigated in some cases. For instance, the effect of the presence of some anions, such as nitrate or carbonate, or natural organic matter (humic acids) on the elimination of clofibric acid by $TiO_2$ and ZnO under sunlight was investigated; while the inorganics played a detrimental role, in the case of humic acids was not so evident and depended on the experimental conditions because of the existence of antagonic effects that will be described in Sect. 5.5 [28]. In other works, tap water has been used in the experiments or the effluent of a wastewater treatment plant has been spiked with relatively high concentrations of selected chemicals. For instance, Xekoukoulotakis et al. studied the elimination of sulfamethoxazole with $TiO_2$ at different pH values and compared the results obtained with three water matrixes: ultrapure water, ground water and treated wastewater, the authors attributed the lower reaction rates measured in ground water and treated wastewater to the aggregation of the $TiO_2$ particles in the presence of larger ion strengths [29]. In the case of photo-Fenton, the presence of the water-matrix composition is also of importance, as some ions, namely chloride act as scavengers of the reactive species.

In this context, the study of the applicability of solar-driven processes for the removal of pharmaceuticals in marine water is of interest, as large amounts of antibiotics are employed in intensive aquiculture. Hence, important volumes of salty water polluted with these chemicals are formed. It has been observed that although elimination of the pharmaceuticals was achieved, the process was less efficient, both using titanium dioxide [16] and photo-Fenton [19], and longer periods of irradiation would be required to decontaminate those effluents.

The treatment of solutions containing mixtures of several pollutants has also been investigated. Although this approach involves the study of a more complex system, it is closer to the real situation. For instance, Klamerth et al. studied the degradation of mixture of 15 emerging pollutants that belong to different families by means of a solar photo-Fenton process [30]: analgesics (acetaminophen, anti-pyrine), herbicides (atrazine, isoproturon), a biocide (hydroxybiphenyl), a stimulant (caffeine), an anticonvulsant (carbamazepine), anti-inflammatory drugs (diclofenac, ibuprofen, ketorolac), antibiotics (flumequine, ofloxacin, sulfamethoxazole and triclosan) and a steroid hormone (progesterone). In another paper, Bernabeu et al. spiked an effluent from a wastewater treatment plant with 5 mg/l of six emerging pollutants and treated with $TiO_2$ in order to obtain accurate kinetics [31]: antibiotics (trimethoprim), analgesic (acetaminophen), anti-inflammatory drug (diclofenac), stimulating drug (caffeine), fungicide (thiabendazole) and a pesticide (acetamiprid). As can be observed, when mixtures are studied, other types of emerging pollutants, such as insecticides, herbicides, personal care products or stimulating agents are also present in addition to pharmaceuticals.

There are some papers reporting on the treatment of emerging pollutants at low concentrations (a few µg/l); in some cases real effluents form wastewater treatment plants have been treated [31]. A more sophisticated analytical methodology is required to deal with those concentrations, as they are close to the quantition limits of conventional techniques: injection of higher volumes in HPLC, pre-concentration of samples by liquid phase extraction or analysis by LC–MS.

The elimination of pharmaceuticals by solar processes, not only has been studied at laboratory scale, but it has also been scaled-up employing pilot plants. The most widely employed plants are based in compound parabolic collectors,

**Fig. 5.4** Picture of a pilot plant for solar detoxification of wastewater based on CPC technology. The plant works in batch mode and it is able to contain 25 l in each process

CPCs (see Fig. 5.4). Briefly, each CPC consist in two parabolic aluminum surfaces which concentrate direct and diffuse radiation in the axis, where is placed a Pyrex glass tube through which the reaction to be treated flows [32]. Bernabeu et al., employed a CPC plant to treat by means of titanium dioxide a real effluent from a wastewater treatment plant [31]. Different emerging pollutants were detected in this effluent in concentration between 0.03 and 15 μg/l: antibiotics (trimethoprim, ofloxacin, enrofloxacin, claritromicin and erythtomycin), analgesic (acetaminophen), anti-inflammatory drugs (diclofenac), psychiatric drugs (carbamazepine), stimulant (caffeine), fungicide (thiabendazole) and pesticides (acetamiprid). The final concentrations were systematically below 50 ng/l at the end of the process; the percentages of elimination depended on the initial concentration of the pollutants, and ranged from the 99% determined for caffeine (whose initial concentration was above 1 μg/l) to the 70% measured for claritrhomycin (33 ng/l at the beginning of the experiment).

Real effluents from a pharmaceutical industry have been treated by means of a solar photo-Fenton process using a pilot plant. The main pollutant was nalidixic acid (ca. 50 mg/l) although the effluent was a complex mixture also containing other organic and inorganic species. Nalidixic was removed after ca. 3 h of irradiation and an increase in the biodegradability of the effluent was observed according to the Zahn-Wellens test. Finally, the possibility of coupling a bioprocess after the photochemical treatment was investigated employing a immobilized biomass reactor; ca. 95% mineralization of the organics was achieved following this approach [25].

However, for real applications with effluents at low concentrations of pollutants exploring treatments at milder conditions appears convenient. Klamerth et al. studied the elimination of a mixture of nine emerging pollutants at pilot plant scale with low amounts of $TiO_2$ (5 mg/l) and photo-Fenton at natural pH values and low iron concentrations (5 mg/l) [30]. Pollutants removal was faster in the case of neutral photo-Fenton than for $TiO_2$; however, this process is limited by the low solubility of iron at neutral pH and by the inefficient formation of hydroxyl radicals under these conditions. It has been observed that the presence of natural organic matter able to form complexes with iron, such as humic acids enhances the efficiency of the process. On the other hand, carbonates and bicarbonates, which are not present in the highly acidic conditions required in conventional photo-Fenton have been described to play a radical scavenging role at neutral pH.

The use of supported titanium dioxide might be advantageous from the practical point of view, as the difficult recovery of the particles of this semiconductor would not be necessary. $TiO_2$ has been immobilized using sintered glass cylinders to eliminate the antibiotic oxolinic acid under black light irradiation [23]. Titania onto borosilicate glass spheres has been employed in a recent work to treat in a solar pilot plant a mixture of 15 emerging pollutants at low concentration (100 µg/l), most of them, pharmaceuticals. Complete removal of twelve pollutants was reached after 50 min of irradiation and only atrazine, carmamazepine and antipyrine remained in the solution; nevertheless, their concentrations were below 50 µg/l. The possibility of using the supported catalyst in different cycles was also demonstrated [33].

Finally, an economic evaluation of a solar photo-Fenton treatment of high concentrations of paracetamol (157 mg/l) in pilot plant has been recently published [24]. They have evaluated that reaction time is the main parameter in the cost of the process; hence, the proposed strategy is to finish the treatment once the effluent is biodegradable enough to be discharged in a biological reactor. Following this procedure a cost of 3.45 €/m$^3$ was estimated.

## 5.5 Photolysis of Pharmaceuticals

A chapter dealing with the use of sunlight for the elimination of pharmaceuticals would not be complete without writing at least a few paragraphs on the self-cleaning of surface waters by solar irradiation. Although it is not actually a method for wastewater treatment, solar photolysis is a common fate for emerging pollutants, and pharmaceuticals in particular, in the environment. Hence, it constitutes an important process for the self-remediation of aquatic natural ecosystems, such as rivers, lakes or seas. These processes are of paramount interest in environmental chemistry although their study is not easy because of the low concentration that pollutants reach in real samples, typically a few ng/l [34] that require sophisticated analytical equipment; additionally, the involvement of transitory species with short lifetimes, whose detection is essential to gain further insight into the fundamentals of these processes, requires application of photophysical measurements.

Direct and indirect mechanisms have been described for the photolysis of pharmaceuticals [3, 35]. In the first case, direct photolysis of the pollutants occurs upon absorption of sunlight. The indirect mechanisms involve generation of highly reactive species, such as hydroxyl radical, superoxide anion or singlet oxygen, which are able to react efficiently with organic matter [36].

Among the species that promote this indirect mechanism are humic acids. They constitute a group of colored substances which are formed by biological processes from vegetal or animal residues; they have been reported to be the major fraction of natural dissolved organic matter in the environment. For instance, it has been found that excited states of natural organic matter promoting removal of amoxicillin under sunlight in aquatic environments [37]. However, humic acid can also absorb light in the UVA-visible range, producing a screen effect on the sample because of light absorption, diminishing direct photolysis; hence, their actual role has to be carefully determined. In this context, Andreozzi et al. found that the presence of humic acids promoted the photo-oxidation of some pharmaceuticals (ofloxacin, sulfamethoxazole, propranolol and clofibic acid) while they inhibited the photo transformation of other drugs (carbamazepine and diclofenac). These authors also reported that some inorganics, such as nitrate, are also able to photogenerate hydroxyl radicals, enhancing the indirect photolysis of pharmaceuticals [38].

Photolytic removal of pollutants can benefit from the synergetic effect of biotic process occurring in the aquatic environment. However, this possibility strongly depends on the toxicity and biodegradability of the parent pollutants and the intermediates formed in the process. As a result of the photolysis of certain chemicals such as sulfamethoxazole [39] an enhancement of the toxicity of the sample was observed according to *V. fischeri* assay, the elimination of the trimethoprim did not result in a significant variation in toxicity [17] and the photolysis of the antibiotic gatifloxacin produced sequential increases and decreases of toxicity, according to the predominating intermediates that are formed after different irradiation periods [40]. Hence, detection of major by-products and elucidation of reaction mechanisms are also important in those cases; however, generally initial concentrations of pollutants higher than the few nanograms detected in surface waters have been employed [17], in order to make detection of the formed intermediates possible.

## 5.6 Concluding Remarks

The ability of titanium dioxide and photo-Fenton to achieve primary removal of pharmaceuticals under solar irradiation is well established. However, most experiments have been carried out at laboratory scale and under experimental conditions, which are far from those of real effluents containing those pollutants.

Hence more effort is required to work with complex mixtures of pollutants with different aqueous matrixes. In addition, it also seems convenient from the technical point of view to work with supported materials in the case of $TiO_2$ and to develop more efficient photo-Fenton processes at mild conditions. Another important issue is to employ analytical techniques (bioassays and chromatographic methods) which permit to rule out the presence of toxic by-products in the treated sample and to gain further insight into the fundamentals of the processes. In conclusion, based on the state of the art the implementation of these processes as tertiary treatments for wastewater treatment plant effluents could be expected in the next future, once the problems described above have been solved.

# References

1. Petrovic M, Barceló D (2006) Liquid chromatography-mass spectrometry in the analysis of emerging environmental contaminants. Anal Bioanal Chem 385:422–424
2. Richardson SD (2008) Environmental mass spectrometry: emerging contaminants and current issues. Anal Chem 80:4373–4402
3. Khetan SK, Collins TJ (2007) Human pharmaceuticals in the aquatic environment: a challenge to green chemistry. Chem Rev 1007:2319–2364
4. Verlicchi P, Galletti A, Petrovic M, Barceló D (2010) Hospital effluents as a source of emerging pollutants: an overview of micropollutants and sustainable treatment options. J Hydrol 389:416–428
5. Ternes TA, Mesisenheimer M, McDowell D, Sacher F, Brauch HJ, Haist-Gulde B, Preuss G, Wilme U, Zulei-Seibert N (2002) Removal of pharmaceuticals during drinking water treatment. Environ Sci Technol 36:3855–3863
6. Murray KE, Thomas SM, Bodour AA (2010) Prioritizing research for trace pollutants and emerging pollutants in the freshwater environment. Envoron Pollut 158:3462–3471
7. Pal A, Yew-Hoong Gin K, Yo-Cen Lin A, Reinhard M (2010) Impacts of emerging organic contaminants in freshwater resources: review of recent occurrences, sources, fate and effects. Sci Total Environ 408:6062–6069
8. Malato S, Fernández-Ibáñez P, Maldonado MI, Blanco J, Gernjak W (2009) Decontamination and disinfection of water by solar photocatalysis: recent overview and trends. Catal Today 147:1–59
9. Gaya UI, Abdullah AH (2008) Heterogeneous photocatalytic degradation of organic contaminants over titanium dioxide: a review of fundamentals, progress and problems. J Photochem Photobiol C Photochem Rev 9:1–12
10. Martínez C, Canle M, Fernández MI, Santaballa JA, Faria J (2011) Kinetics and mechanism of aqueous degradation of carbamazepine by heterogeneous photocatalysis using nanocrystalline $TiO_2$, ZnO, and multiwalled carbon nanotubes-anatase composite. Appl Catal B Environ 102:563–571
11. Palominos RA, Mondaca MA, Giraldo A, Peñuela G, Pérez-Moya M, Mansilla HD (2009) Photocatalytic oxidation of the antibiotic tetracycline on $TiO_2$ and ZnO suspensions. Catal Today 144:100–105
12. Pignatello JJ, Oliveros E, Mackay A (2006) Advanced oxidation processes for organic Contaminant destruction based on the Fenton reaction and related chemistry. Critical Rev Environ Sci Technol 36:1–84
13. An T, Yang H, Li G, Song W, Cooper WJ, Nie X (2010) Kinetics and mechanism of advanced oxidation processes (AOPs) in degradation of ciprofloxacin in water. Appl Catal B 94:288–294

14. An T, Yang H, Li G, Song W, Luo H, Cooper WJ (2010) Mechanistic considerations for the advanced oxidation treatment of fluoroquinolone pharmaceutical compounds using $TiO_2$ heterogeneous catalysis. J Phys Chem A 114:2569–2575

15. Hapeshi E, Achilleos A, Vasquez MI, Michael C, Xekoukoulotakis NP, Mantzavinos D, Kassinos D (2010) Drugs degrading photocatalytically: kinetics and mechanisms of ofloxacin and atenolol removal on titania suspensions. Water Res 44:1737–1746

16. Sirtori C, Agüera A, Gernjak W, Malato S (2010) Effect of water-matrix composition on trimethoprim solar photodegradation kinetics and pathways. Water Res 44: 2735–2744

17. Piecha M, Sarakha M, Trebse P (2010) Photocatalytic degradation of cholesterol-lowering statin drugs by $TiO_2$-based catalyst. Kinetics, analytical studies and toxicity evaluation. J. Photochem Photobiol A Chem 213:61–69

18. Elmolla ES, Chaudhuri M (2011) The feasibility of using combined $TiO_2$ photocatalysis-SBR process for antibiotic wastewater treatment. Desalination 272:218–224

19. Trovó AG, Nogueira RFP, Agüera A, Fernández-Alba AR, Sirtori C, Malato S (2009) Degradation of sulphamethoxazole in water by solar photo-Fenton. Chemical and toxicological evaluation. Water Res 43:3622–3931

20. Romero V, De la Cruz N, Dantas RF, Marco P, Gimenez J, Esplugas S (2011) Photocatalytic treatment of metropolol and propranolol. Catal Today 161:115–120

21. Zhao Ch, Deng H, Li Y, Liu Z (2010) Photodegradation of oxytetracicline in aqueous by 5A and 13X loaded with $TiO_2$ under UV irradiation. J Hazard Mater 176:884–892

22. Achilleos A, Hapeshi E, Xekoukoulotakis N, Mantzavinos D, Fatta-Kassinos D (2010) Factors affecting diclofenac decomposition in water by UV-A/$TiO_2$ photocatalysis. Chem Eng J 161:53–59

23. Palominos RA, Mora A, Mondaca MA, Pérez-Moya M, Mansilla HD (2008) Oxolinic acid photo-oxidation using immobilized $TiO_2$. J Hazard Mater 158:460–464

24. Santos-Juanes L, Ballesteros MM, Ortega E, Cabrera A, Román IM, Casas JL, Sánchez JA (2011) Economic evaluation of the photo-Fenton process. Mineralization level and reaction time: the keys for increasing plant efficiency. J Hazard Mater 186:1924–1929

25. Sirtori C, Zapata A, Oller I, Gernjak W, Agüera A, Malato S (2009) Decontamination of industrial pharmaceutical wastewater by combining solar photo-Fenton and biological treatment. Water Res 43:661–668

26. Nieto J, Freer J, Contreras D, Candal RJ, Sileo EE, Mansilla HD (2008) Photocatalyzed degradation of flumequine by doped $TiO_2$ and simulated solar light. J Hazard Mater 155: 45–50

27. Rozas O, Contreras D, Mondaca MA, Pérez-Moya M, Mansilla HD (2010) Experimental design of Fenton and photo-Fenton reactions for the treatment of ampicillin solutions. J Hazard Mater 177:1024–1030

28. Li W, Lu S, Qiu Z, Lin K (2011) Photocatalysis of clofibric acid under solar light in summer and winter seasons. Ind Eng Chem Res 50:5384–5393

29. Xekoukoulotakis NP, Drosou C, Brebou C, Chetzisymeon E, Hepeshi E, Fatta-Kassinos D, Mantzavinos D (2011) Kinetics of UV-A-$TiO_2$ photocatalytic degradation and mineralization of the antibiotic sulfamethoxazole in aqueous matrices. Catal Today 161:163–168

30. Klamerth N, Rizzo L, Malato S, Maldonado MI, Agüera A, Fernández-Alba AR (2010) Degradation of fifteen emerging contaminants at µg/L initial concentrations by mild solar photo-Fenton in MWTP effluents. Water Res 44:545–554

31. Bernabeu A, Vercher RF, Santos-Juanes L, Simón PJ, Lardín C, Martínez MA, Vicente JA, González R, Llosá C, Arques A, Amat AM (2011) Solar photocatalysis as a tertiary treatment to remove emerging pollutants from wastewater treatment plant effluents. Catal Today 161:233–240

32. Malato S, Blanco J, Vidal A, Richter C (2002) Photocatalysis with solar energy at a pilot-plant scale: an overview. Appl Catal B Environ 37:1–15

33. Miranda-García N, Maldonado MI, Coronado JM, Malato S (2010) Degradation study of 15 emerging contaminants at low concentration by immobilized $TiO_2$ in pilot plant. Catal Today 151:107–113

34. Ziylan A, Ince NH (2011) The occurrence and fate of anti-inflamatory and analgesic pharmaceuticals in sewage and fresh water: treatability by conventional and non-conventional processes. J Hazard Mater 187:24–36
35. Tixier C, Singer HP, Oellers S, Müller SR (2003) Occurrence and fate of carbamazepine, clofibric acid, diclofenac, ibuprofen, ketoprofen, and naproxen in surface waters. Environ Sci Technol 37:1061–1068
36. Razavi B, Ben AS, Song W, O'Shea KE, Cooper WJ (2011) Photochemical fate of atorvastatin (lipitor) in simulated natural waters. Water Res 45:625–631
37. Xu H, Cooper WJ, Jung J, Song W (2011) Photosensitized degradation of amoxicillin in natural organic matter isolate solutions. Water Res 45:632–638
38. Andreozzi R, Raffaele M, Nicklas P (2003) Pharmaceuticals in STP effluents and their solar photodegradation in aquatic environment. Chemosphere 50:1319–1330
39. Trovó AG, Nogueira RF, Agüera A, Sirtori C, Fernández-Alba AR (2009) Photodegradation of sulfamethoxazole in various aqueous media: persistence, toxicity and photoproducts assessment. Chemosphere 77:1292–1298
40. Ge LK, Chen JW, Zhang SY, Cai XY, Wang Z, Wang CL (2010) Photodegradation of fluoroquinolone antibiotic gatifloxacin in aqueous solutions. Chin Sci Bull 155:1495–1500

# Chapter 6
# Outlook

**Giusy Lofrano**

Significant progress has been made in the recent decades in recognising and understanding the issues in sustainability. The rate of population growth, the level of economic development, often equated with quality of life, and environmental protection have long been recognised challenges to mutually create a sustainable future. Historical evidences proved that an increasing human population has put an increasing demand on natural resources used for consumption and waste management. The challenge of green engineering decouples the historical relationship of population growth and environmental degradation on the path towards sustainability that means an improved quality of life.

Among several green technologies applied to wastewater treatment for emerging compounds removal, this book focuses on natural (adsorption and constructed wetlands) and advanced solar-based treatments because their characteristics make them inherently green.

Many studies of non-conventional treatments are available in the literature; nevertheless, they are often limited to laboratory scale. Indeed, the cost estimation of adsorption processes utilising low-cost adsorbents is not strictly right and pilot-plant studies should also be carried out to check their feasibility on commercial scale.

Interesting results on removal of organic micropollutants, particularly Pharmaceutical and Personal Care Products (PPCPs), came out from constructed wetlands. Their application to small communities or as tertiary treatments dealing with a small, diverted fraction of conventional effluents from Wastewater Treatment Plants (WWTP) appears quite attractive both due to the limited energy

G. Lofrano (✉)
Department of Civil Engineering, University of Salerno,
via ponte don Melillo, 84084 Fisciano (SA), Italy
e-mail: glofrano@unisa.it

G. Lofrano (ed.), *Emerging Compounds Removal from Wastewater*,
SpringerBriefs in Green Chemistry for Sustainability,
DOI: 10.1007/978-94-007-3916-1_6, © Lofrano 2012

required and to the relatively low maintenance costs, which contribute to make this technology a unique green technology.

From the studies provided in this book it can be seen that solar-advanced oxidation processes are effective treatment methods for the removal of trace pollutants. However, there are a number of issues to be solved pertaining these treatment methods, involving the identification of the oxidation by-products as well as intermediates, the evaluation of biodegradability, and potential estrogenic activity of these compounds. Furthermore, as the process costs may be considered the main obstacle to their commercial application, several promising cost-cutting approaches have been proposed, such as integration of Advanced Oxidation Processes AOPs as a part of a treatment train.

Even more in the near future, green chemistry should focus on the development of economically feasible conversion of solar energy into chemical energy and improvement in the conversion of solar energy to electric power.

Testing sustainability requires to think long and hard. It is the time to begin.